Phylogenetic Biology

Phylogenetic Biology

First Edition

Version 1.0.0

Casey W. Dunn

https://dunnlab.org

Published by Casey W. Dunn

New Haven, CT, USA

ISBN 979-8-9934524-0-1 (paperback)

ISBN 979-8-9934524-1-8 (hardcover)

Library of Congress Control Number: 2026918096

DOI: https://doi.org/10.5281/zenodo.17267993

Cover art: *Moth Cluster IV* (2016), pen, ink and silkscreen on paper, 65 × 140 inches. © James Prosek; image courtesy of the artist.

Cover design by Casey W. Dunn.

This book is dedicated to my sister Jenny Dunn, who has been there every step of the way. I am so grateful for her adventurous spirit, quick mind, and love of life.

Contents

Preface **xi**
Approach xi
Using this book xi
Distribution xi
Other resources xii
Acknowledgments xiii

1 Introduction **1**
1.1 History of the field 1
1.2 Core concepts 2
1.3 A unified perspective on phylogenetic studies 3
1.4 Applications 6

2 Phylogenies **9**
2.1 Phylogenies are graphs 10
2.2 Drawing phylogenies 12
2.3 Some properties of phylogenies 14
2.4 Rotating nodes 15
2.5 The meaning of branch lengths 15
2.6 Naming nodes and groups 17
2.7 Polytomies 19
2.8 Rooting 19
2.9 The information contained in phylogenies 24
2.10 Representation 26
2.11 Additional reading 29

3 Simulation **31**
3.1 Models 31
3.2 A simple model of evolution 34
3.3 Generalizing the simple model 45
3.4 Expanding the models 47
3.5 Plugging some numbers into the expanded model 49
3.6 Simulation along a single branch 51
3.7 Scaling from a single branch to a whole tree 52
3.8 Scaling from a single site to multiple sites 54
3.9 Concluding thoughts 54

3.10 Additional resources 54

4 Inferring phylogenies from data **55**
4.1 Probability of a single history 55
4.2 Probability of multiple histories 58
4.3 Log likelihood 61
4.4 Likelihood for multiple sites 62
4.5 Maximum likelihood 62
4.6 Optimality criteria 63

5 Molecular inference in practice **65**
5.1 Your question 65
5.2 Taxon sampling 65
5.3 Collecting sequence data 67
5.4 Data processing upstream of phylogenetic analyses 70
5.5 Working with publicly available data 71
5.6 Homologous sites 73
5.7 Combining information across genes 73

6 Evaluation **75**
6.1 Model evaluation 75
6.2 Topological evaluation 80
6.3 Topology tests 83

7 Gene trees and species trees **85**
7.1 Gene trees, species trees, and real discordance 85
7.2 Incomplete lineage sorting 86
7.3 Gene duplication and loss 89
7.4 Reticulation: introgression, hybridization, and horizontal transfer 91
7.5 Conclusion 91

8 Bayesian phylogenetics **93**
8.1 Bayesian statistics 94
8.2 Bayesian phylogenetic inference 95
8.3 Bayesian phylogenetic inference in practice 97
8.4 Resources 99

9 Additional character types **101**
9.1 Discrete character types 103
9.2 Continuous data 105

10 Phylogenies and time **109**
10.1 Measurements of time on trees 109
10.2 Time calibration 111
10.3 The implications of constraining node ages 112

11 Ancestral character state reconstruction **115**
11.1 Reconstructing discrete characters 116
11.2 Reconstructing continuous characters 118

11.3 Cautions and limitations 120

12 Comparative methods **121**
12.1 Species are not independent observations 121
12.2 Phylogenetic independent contrasts 123
12.3 The expected covariance structure 126
12.4 Phylogenetic generalized least squares 127
12.5 Beyond correlated evolution 128

A Statistical and mathematical fundamentals **131**
A.1 Probability theory and general statistics 131
A.2 Bayes' theorem 131
A.3 Linear algebra 132

B Software versions **133**

Bibliography **137**

Index **147**

Preface

Approach

Models are at the core of modern phylogenetic biology. Models are usually taught in the context of phylogenetic inference, where they are used to look backward in time. This seems to me a bit like teaching students to master driving in reverse before you show them how to drive forward down a road. Phylogenetic models are generative—they explain how data are created under specific evolutionary processes. They are therefore more intuitive to understand in the context of generating data, proceeding in time from ancestral states forward to future states. I therefore focus first on building an intuitive understanding of models in the context of simulation, and only after that apply models to inference and other tasks that look back in time.

This book is not a comprehensive treatment of phylogenetic biology; it is focused on teaching some of the key concepts and methods. This will provide a foundation for the many other rich areas of the field, from its interfaces with population genetics to comparative genomics to comparative methods at a macroevolutionary scale. It is in no way an overview of all of phylogenetic biology, which has grown into a very large field. Omissions of topics do not imply that they are unimportant.

Using this book

This book is intended both for the self-directed learner and for use in a course.

I wrote it as a text for my course, Phylogenetic Biology (Yale EEB354). We read one chapter a week. We review and discuss the reading on Tuesdays, and then on Thursdays do hands-on work, read papers from the literature, or share student projects.

Distribution

There are several ways you can get this book:

- You can read an HTML version for free at https://dunnlab.org/phylogenetic_biology/

- You can purchase a paperback from your favorite book retailer. Search for ISBN 979-8-9934524-0-1.
- You can purchase a hardcover from your favorite book retailer. Search for ISBN 979-8-9934524-1-8.
- The book is rendered from source code available at https://github.com/caseywdunn/phylogenetic_biology, with `bookdown` (Xie, 2016). If you are curious about how any of the figures or analyses were done you can examine the source code there and rerun it yourself.
- Snapshots of source code for each release are also available on Zenodo via the book's DOI at https://doi.org/10.5281/zenodo.17267993.

Please submit any errors you find, typos, or suggestions that you have for improving the manuscript to the issue tracker at https://github.com/caseywdunn/phylogenetic_biology/issues.

You are currently reading the First Edition (Version 1.0.0) of the book. Please cite it as follows:

> Dunn CW. Phylogenetic Biology. First Edition. New Haven, CT: Casey W. Dunn; 2026. ISBN 979-8-9934524-0-1. https://doi.org/10.5281/zenodo.17267993

This book follows semantic versioning (major.minor.patch). The major version is the edition: a new edition marks a substantial revision and carries a new ISBN, minor versions add or revise content while keeping earlier citations valid, and patch versions make editorial corrections. The online edition at https://dunnlab.org/phylogenetic_biology/ is updated continuously as new versions are released, while each released version is permanently archived on Zenodo through the DOI above.

Other resources

The following sites have a wide variety of material that is relevant to the theory and practice of phylogenetic biology.

- An extensive list of tools, tutorials, and examples of phylogenetic tools in the programming language R maintained by Brian O'Meara. https://cran.r-project.org/web/views/Phylogenetics.html.
- Liam Revell has an excellent blog on phylogenetic methods. https://blog.phytools.org/.
- The Workshop on Molecular Evolution at Woods Hole. This is an intensive summer course on phylogenetics, with an emphasis on building phylogenetic trees. https://molevolworkshop.github.io.
- The Applied Phylogenetics Workshop in Bodega Bay. This is another summer course on phylogenetics, but with a bit more emphasis on using phylogenies to test evolutionary questions. http://treethinkers.org/tutorials/.

The following are some of the many great books for learning more about phylogenetic methods:

- Baum, D. and Smith, S. (2012). Tree Thinking: An Introduction to Phylogenetic Biology. Roberts Publishers.
- Felsenstein, J. (2004). Inferring Phylogenies. Sinauer Associates.
- Garamszegi, L. Z. (2014). Modern Phylogenetic Comparative Methods and Their Application in Evolutionary Biology. Springer.
- Paradis, E. (2011). Analysis of Phylogenetics and Evolution with R. Springer.
- Revell, L. J. and Harmon, L. J. (2022). Phylogenetic Comparative Methods in R. Princeton.
- Swofford, D. L., Olsen, G. J., Waddell, P. J., and Hillis, D. M. (1996). Phylogenetic inference. In: Molecular Systematics, Second Edition. Eds: D. M. Hillis, C. Moritz, and B. K. Mable. Sinauer Associates.

The following books provide general computational background for the topics covered here:

- Wickham, H. and Grolemund, G. (2017). R for Data Science. https://r4ds.had.co.nz.
- Haddock, S. H. D. and Dunn, C. W. (2010). Practical Computing for Biologists. http://practicalcomputing.org.

Acknowledgments

Thanks in particular to the students of Yale EEB354 in 2020, 2022, and 2024. This book started as a collection of lecture notes for this course. The students provided invaluable motivation and feedback. Thanks in particular to Lauren Mellenthin (graduate teaching fellow for the course in 2020), Namrata Ahuja (teaching fellow in 2022), and Dalila Destanovic (teaching fellow in 2024). Other lab members provided very helpful feedback when I posted new chapters. Steve Haddock and Felipe Zapata also provided close reads of most chapters, often within hours of completing first drafts. Nate Grubaugh invited me to share a couple chapters in his course each year. Thanks to Richard Hammack, author of Book of Proof, for his helpful advice on self-publishing to facilitate student access. I am also very grateful to the students and faculty of the Workshop on Molecular Evolution at Woods Hole. Finally, I thank James Prosek for generously allowing me to use his work *Moth Cluster IV* for the cover.

Chapter 1

Introduction

Phylogenetic biology is the study of evolutionary relationships, and the use of those relationships to study other aspects of biology.

1.1 History of the field

At the dawn of the field, in the second half of the 19th century (Darwin, 1859; Haeckel, 1897), phylogenetic biology was largely a speculative endeavor. People looked at similarities and differences across organisms, and made hypotheses about the evolutionary relationships that would give rise to those patterns. It wasn't until the second half of the 20th century that explicit methods were developed to infer phylogenetic relationships (Hennig, 1966). This ushered in decades of accelerating phylogenetic methods development that continues to this day. The introduction of explicit model-based approaches to phylogenetic inference was particularly influential (Felsenstein, 1981). These new methods were computationally intensive, but fortunately their rise paralleled the rapid development and widespread availability of powerful computers.

Beginning in the late 1980s, molecular data became widely used, based on sequence fragments of a few genes, for building phylogenies of extant organisms. Starting in about 2010, new generations of high-throughput sequencing technology made it possible to collect sequences for thousands of genes from a broad diversity of organisms in a single study (Lemmon and Lemmon, 2013), and right now we are in the earliest days of building phylogenies from broadly sampled high-quality chromosome-level genome assemblies.

Phylogenetic biology has always been concerned with figuring out evolutionary relationships. This domain of inquiry is referred to as phylogenetic inference. Phylogenetic inference has been vital to improving the classification of species (phylogenetic taxonomy) (de Queiroz, 2007) and to reconstructing key features in the evolutionary history of many groups of organisms, and it is of deep interest to those who know the organisms well.

Many questions in biology require knowing evolutionary relationships, even if those evolutionary relationships are themselves not of central interest to the investigator. This has led to the rise of another domain often referred to as phylogenetic comparative biology (Felsenstein, 1985b). Instead of ending with trees, as many phylogenetic inference projects do, phylogenetic comparative projects usually start with trees and use them to study the evolution of traits. Comparative questions include whether there is evolutionary covariance between traits, or shifts in rates of evolution along particular branches (also referred to as edges). Phylogenetic comparative methods have become an increasingly large part of the field.

Though phylogeny initially sprang from the field of comparative morphology, and much of the initial focus was on building and analyzing trees with morphological data and using them to study the evolution of morphology, the field has rapidly expanded to encompass many other categories of data. Even so, morphology continues to be vital to building phylogenies in many contexts, including when fossils are available. While most phylogenetic inference is now based on molecular data, phylogenetic comparative methods are routinely applied to all sorts of data and questions. Phylogenetic comparative methods now play important roles in the study of physiology, ecology, genomics, medicine, and most other parts of biology.

Phylogenetics is no longer a strictly macroevolutionary field. Some of the most exciting work in recent years has been at the interface of population genetics and phylogenetic biology, helping to unify our perspectives on micro- and macroevolution. Phylogenies are now also routinely used to study the evolution of genes and other molecular characters, and also to dive into extremely recent and fast evolutionary processes within species, such as viral pandemics.

1.2 Core concepts

There are a few concepts that are fundamental to understanding phylogenetic analyses (Figure 1.1A). These include:

- The topology of the phylogeny. This is the structure of the evolutionary relationships. It can be thought of as the branching order of a tree.
- The lengths of the branches in the phylogeny. Length can signify different things, such as time elapsed or amount of expected evolutionary change.
- Characters. The organism attributes under consideration. These include things like leg length, geographic location, a particular site in a particular gene, a physiological attribute, or a feature of genome organization. Characters are also often called traits.
- Character states. The particular values that can be taken by different individuals for specific characters. If leg length is a character, a leg length character state could be 4.3 mm. For a site in a DNA sequence, a character state could be `C`, `G`, `T`, or `A`.

- Models. These are hypotheses about how characters evolve. They take a mathematical and statistical form. For example, a model could describe the covariance between evolutionary changes in different characters, or be a set of functions that describe the probabilities of specific state changes along a branch. Models are abstract representations of biological processes. They are deliberate simplifications that allow us to explicitly describe what we think the most important features of change are. We can make models as simple or complex as we like, but the more complex they are the more information we need to use them effectively.
- Model parameter values. These are the specific values for terms in the model. In a model that indicates covariance between characters, the specific values in the covariance matrix are model parameter values. If you have a set of functions that describe the probability of changing between particular states for a character, the model parameters include the rates in those functions.

In order to make informative comparisons across species, both for inferring phylogenies and for phylogenetic comparative analyses, we need a unified ontological framework to refer to the same characters and character states in different organisms. This correspondence is provided by the concept of homology. Homology is a hypothesis that the same attribute is present in different entities because that attribute was present in their shared ancestor. Human arms and bird wings are homologous, for example. Homology is a deceptively simple concept that can sometimes be devilishly difficult to define, test, and apply (Wagner, 2014).

1.3 A unified perspective on phylogenetic studies

Phylogenetics is such a diverse and quickly growing field that sometimes studies can seem to be more different from each other than they are. This is a shame, as it is a lost opportunity to share ideas across domains in the field and to apply what is learned in one context to another context.

Fortunately, phylogenetic biology has a strong conceptual foundation that provides a unified perspective on phylogenetic studies (Höhna et al., 2014). Once you have this in mind, you can better see the relationships between what may initially appear to be very different methods, questions, and analyses.

Consider some of the concepts we introduced above:

- Topology
- Branch lengths
- Characters
- Character states of tips
- Character states of ancestors
- Model
- Model parameters

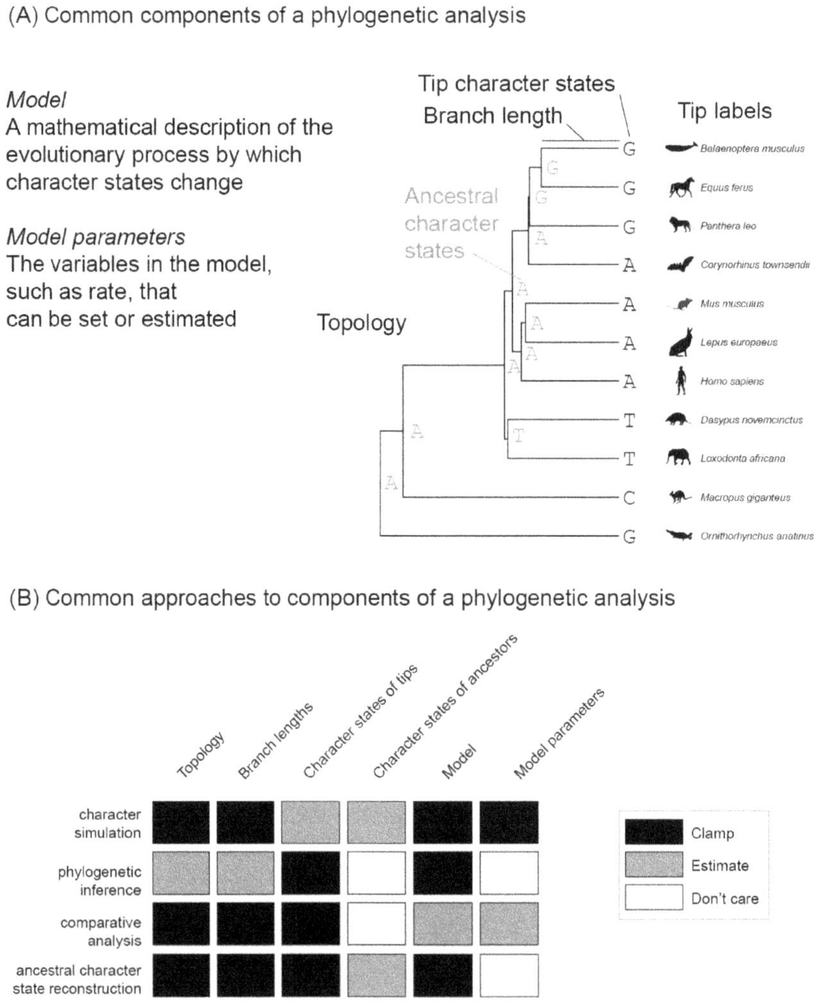

Figure 1.1: (A) The primary components of a phylogenetic analysis. This mammal phylogeny and branch lengths are from http://vertlife.org. The organism silhouettes are from http://phylopic.org/. (B) Different analyses tend to take different approaches to these components.

The investigator can generally take one of three approaches to each of these (Figure 1.1B). They either:

- Clamp the value according to information in hand, such as their own data or the results of a previous study
- Estimate the value
- Don't care about the value

Many phylogenetic inference studies use expert knowledge to define and clamp the characters, run some preliminary analyses to evaluate and then clamp the model, and clamp the character states at the tips according to the observed data such as gene sequences. They then estimate the topology, branch lengths, character states of ancestors, and model parameters. They then throw away the estimates of ancestral character states and model parameters, and present the topology and branch lengths as the result. In the end you get a tree of the inferred evolutionary relationships between your organisms of interest.

Many phylogenetic comparative analyses take the same approach to characters and observed character states at the tip (both clamped), and ancestral character states (estimated, but discarded). They then take a different approach to the remaining items. They clamp the tree and branch lengths according to the results of an earlier phylogenetic inference, evaluate different models, and estimate model parameters. The models and model parameters are then the presented results. They would report, for example, that a model that allows for shifts in rates of evolution along different branches is a better fit than a model that doesn't, or provide estimates of the evolutionary covariance between traits.

Other studies apply the three approaches (clamp, estimate, don't care) in different ways across the phylogenetic study. Some combinations are very common, others very rare. Some have yet to be explored at all. There are a few reasons for this variation. One is historical. Some combinations became widespread early on and stayed that way as later studies were created in the image of earlier studies. Another reason is methodological. Some combinations are rare or nonexistent in practice because they require methods development that hasn't been done yet. Data limitations are another big issue. It takes more data to estimate more things. As you get more data, you can unclamp more things. And the biggest limitation is often computational. Phylogenetic analyses are often computationally expensive, and this can quickly become limiting. This is often the ceiling in practice, but more efficient software tools, improved methods, and new data are all helping to make analyses more tractable at the same time that more computer power is becoming available to more investigators.

Once you dig in, the "don't care" case is surprisingly interesting. Sometimes if you don't care about something, you don't need to estimate it. But often you do need to estimate something intermediate to get to what you do care about, and then you just throw away the intermediate result. Things that you aren't specifically interested in but that you need to estimate to figure out what you do want to learn are referred to as nuisance parameters. Some nuisance parameters aren't even retained by software, but others are. I mean "don't care" in the specific sense that these entities aren't part of your primary question. You should very much care whether your nuisance parameter estimates are sensible since other results depend on them, and you should examine them to make sure they are sensible.

Some studies differ only in which estimates are kept and which are thrown away. One person's nuisance is another's bread and butter. One study may keep ancestral character states and discard model parameters, and another study may run essentially the same analysis but discard ancestral character states and keep model parameters.

1.4 Applications

With this framework in mind, let's take a look at a sampling of recent phylogenetic studies.

- Nextstrain provides a frequently updated phylogeny of SARS-CoV-2, with associated data on geography, sampling time, and other factors (Hadfield et al., 2018). This supported genomic epidemiology, surveillance, and public-health decisions during the SARS-CoV-2 pandemic. They are using observed character data at the tips (virus genome sequences) to estimate topology, branch lengths, and model parameters.
- Myxozoa are enigmatic parasites that infect fish and invertebrate hosts. They have very few morphological traits that show clear homology with other animals. So Sally Chang, Paulyn Cartwright, and colleagues used phylogenetic inference to examine their relationships to other animals, and found strong support for their placement within Cnidaria (Chang et al., 2015).
- My own lab wanted to see if the evolution of gene expression is more rapid after gene duplication than after speciation (Dunn et al., 2018). We found no evidence for such a difference. We used the results of previous studies to clamp gene phylogenies and gene expression character states at the tips, and estimated ancestral expression states to examine shifts in expression along particular branches.
- *Viburnum* is a group of plants with wide distribution in many different habitat types. Michael Landis spearheaded a collaborative project between the Edwards and Donoghue labs here at Yale EEB to understand how *Viburnum* first diversified in Asia during the Eocene and then spread across the globe (Landis et al., 2021). They clamped the character states at the tips based on gene sequence data, morphology, and geographic range, and developed innovative methods to simultaneously estimate the phylogeny, branch lengths, and ancestral geographic ranges. This is the vanguard of new approaches that simultaneously consider multiple categories of data rather than bolt together multiple independent methods that estimate one thing at a time.

This is a very exciting time in phylogenetic biology. For many years most studies followed a few basic templates. With the development of new phylogenetic methods, new tools to collect high-throughput character data, and a growing interest in phylogenetic questions, the field is now in its most interesting stage of development and application yet.

Chapter 2

Phylogenies

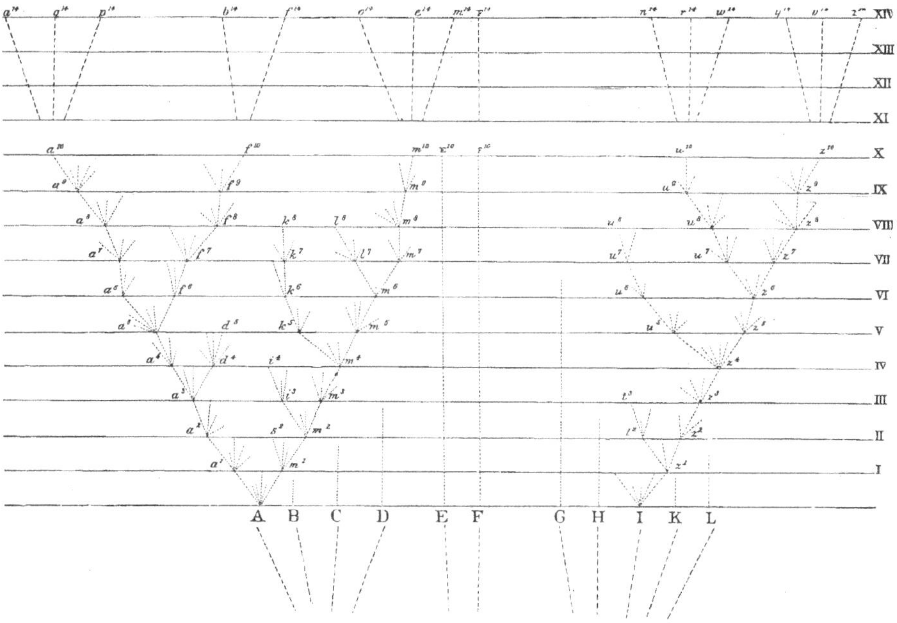

Figure 2.1: Darwin's depiction of the evolutionary relationships among organisms (Darwin, 1859).

Phylogenies represent evolutionary relationships. The only figure in Darwin's *Origin of Species* (Darwin, 1859) was a phylogeny (Figure 2.1), though he didn't call it that.

Phylogenies are often referred to as trees. It is an apt analogy. From the anchoring of the tree at a root, to the depiction of lineages as branches, to the presentation of tips as leaves, there is a direct correspondence. Many of the first speculative phylogenies were illustrated as actual trees, including Haeckel's tree

(Figure 2.2) (Haeckel, 1897). Haeckel had coined the term *phylogeny* decades earlier, in his 1866 *Generelle Morphologie der Organismen* (Haeckel, 1866).

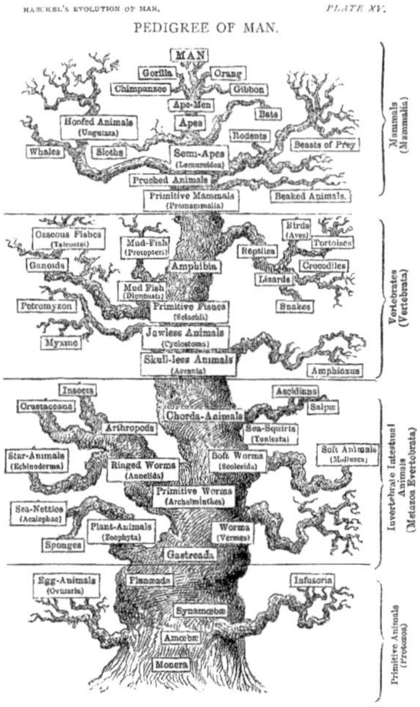

Figure 2.2: Haeckel's hypothesis of the animal phylogeny (Haeckel, 1897), drawn as an actual tree.

Not all trees in biology are phylogenies, though. For example, hierarchical clustering of traits across species results in trees that represent similarity of species, but does not necessarily show evolutionary relationships. In some cases phylogeny and species clustering will be largely congruent, but in many cases they will not be. Hierarchical tree-like clustering is also used for things that have no evolutionary component, like similarity of responses to a drug.

It is helpful, then, to be more precise about the components of a phylogeny, and what each of those parts represents.

2.1 Phylogenies are graphs

In order to establish a strong foundation for thinking about phylogenies, it is helpful to turn to math.

A phylogeny is a specific instance of a mathematical object known as a graph (West, 2001). A graph consists of nodes (also called vertices), often represented as circles, connected by edges, often represented by lines (Figure 2.3). Graphs are useful representations of a variety of systems. For example, nodes could be people, and edges could represent family relationships, connections in a social

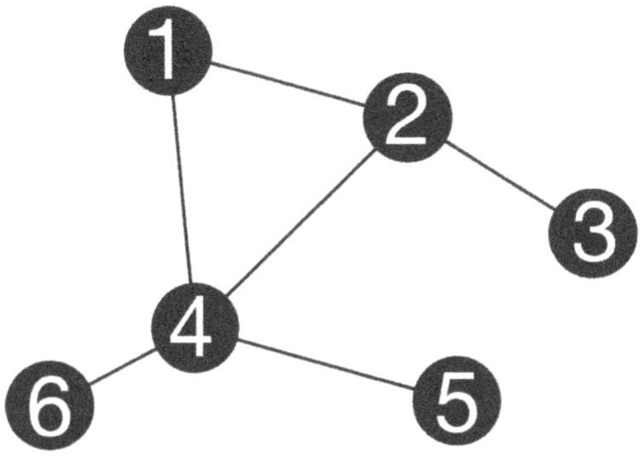

Figure 2.3: Simple graph. Nodes are represented by circles, and edges by lines. Note that this graph has a cycle (you could walk in a circle between nodes 1, 2, and 4). It is also not bifurcating.

network, or financial transactions. We often refer to the edges in phylogenies as branches.

Phylogeny graphs (Figure 2.4) usually have a few specific topological properties. The mathematical properties of phylogenetic trees, including the many results that follow from these properties, are treated in detail by Semple and Steel (2003).

- They are acyclic. This means that there is only one possible path along edges from one node to another. It isn't possible to go in circles. Exceptions can arise, for example in cases of horizontal gene transfer or hybridization.
- They tend to be bifurcating. This means that each internal node has one parent node and two daughter nodes, and therefore three branches connected to it. This represents the biological understanding of speciation, which usually proceeds by one lineage giving rise to two lineages.
- They can be directed. This means that branches are not symmetric, and instead have directionality. Specifically, they denote time. The node on one end of the branch is older than the node on the other end of the branch. We therefore can refer to the nodes connected to a particular branch as the parent node and child node.

A phylogenetic graph is an abstraction, and for it to be useful it is important to keep in mind what features of biology we are attempting to represent. The nodes are entities that can evolve, like organisms or genes. The branches indicate evolutionary relationships between those entities. You could imagine, as an extreme case, a graph that showed every single individual that ever existed in your group of interest, say mammals. Each branch would connect literal parents and offspring. That would be a big phylogeny, and you would never have enough information to know it all, but it does exist even if unknowable and unwieldy given our current tools. A phylogeny is a subset of that graph, where we often retain a single individual per species as the tip nodes, and retain nodes

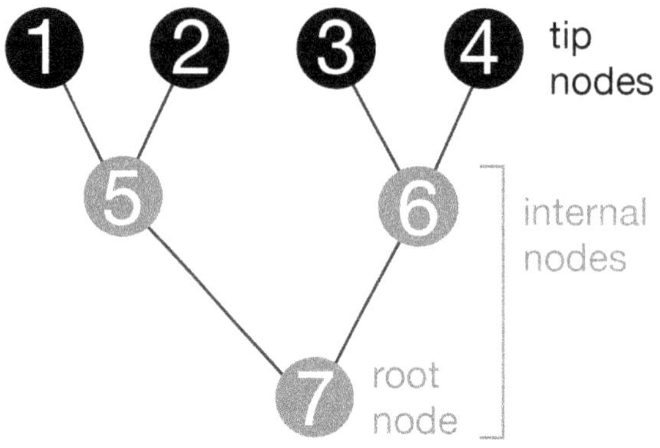

Figure 2.4: Graph representing a phylogeny. The tip nodes typically represent sampled entities, like living organisms, sequenced genes, or fossils. The internal nodes are ancestors that immediately precede divergence events. The root is a special internal node that has no parent. It is acyclic (i.e., there are no paths that go in circles). The graph is bifurcating—the root node has two branches that connect it to its children; the other internal nodes each have three branches (one that connects to the parent, two to children); and the tips each have one branch (that connects to its parent).

immediately preceding divergence events as the internal nodes. In this respect, a phylogeny is a subgraph of the entire history of life on Earth. Because many generations are collapsed along a single branch, parent-child node relationships in the tree don't mean actual parent-child relationships. Parent-child nodes are often separated by millions of generations.

2.2 Drawing phylogenies

Here we will work with some mammal phylogenies, mostly because their silhouette icons are so recognizable and because there are extensive phylogenies available for this group (Upham et al., 2019).

You will rarely see a phylogeny depicted as in Figure 2.4, with a circle for every node. It gets too cluttered. Instead, it is simpler to draw just the branches (Figure 2.5), with the nodes implied at the ends of branches.

There is considerable variation in how the same phylogeny can be drawn (Figure 2.6). This gives some flexibility in deciding what presentation is best for your particular goals, but can also create confusion because two images that look very different may in fact be of the exact same phylogeny.

Rectangular layouts are the most common, because the entire branch length is along one axis of the plot. In a rectangular tree, each node is depicted as a line that is perpendicular to the branches. If the branches are horizontal, so that time reads from left to right, these node lines are vertical lines at the end of each internal branch (Figure 2.6A). The confusing thing is that, because these node lines have the same width and color as the branches, it looks as if they are

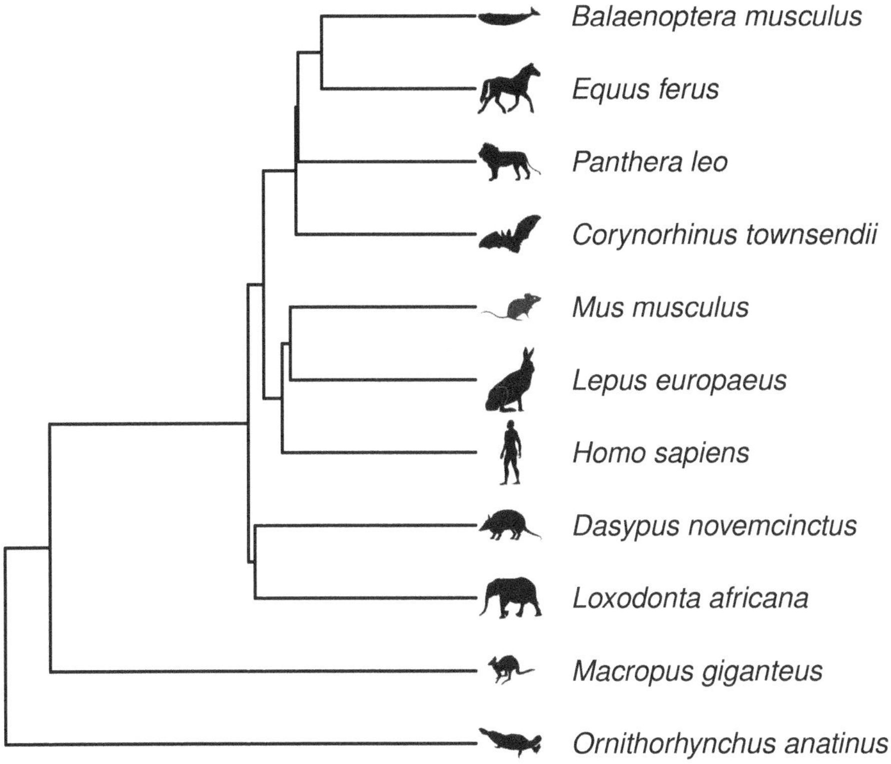

Figure 2.5: Phylogeny of some mammals. Topology and branch lengths from http://vertlife.org. The organism silhouettes are from http://phylopic.org/. Note that node circles are not drawn; instead, nodes are implied at ends of branches.

part of the branches. They aren't though—their length is arbitrary, and they just space out the daughter branches. Node lines also add right-angle elbows where the ends of the node lines connect to the branches, forming a corner.

Slanted layouts avoid the node lines and elbows of rectangular layouts, but because the branches aren't parallel and can be at a variety of angles, it isn't always easy to read branch lengths on them.

Circular layouts place the root at the center, and the tips around the circumference. This is a compact representation that works well for very large phylogenies.

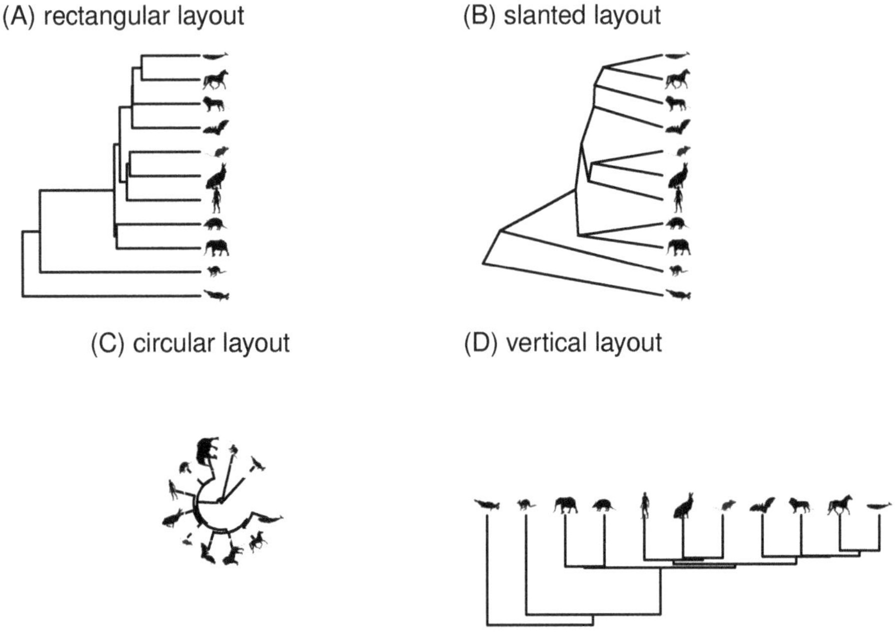

Figure 2.6: The same tree as above, laid out in several different ways.

2.3 Some properties of phylogenies

In a fully bifurcating phylogeny with n tip nodes, the following will be the case:

- There are $n-1$ internal nodes. For example, Figure 2.4 shows a phylogeny with 4 tip nodes (labeled 1-4) and 3 internal nodes (labeled 5-7). The total number of nodes (tip nodes and internal nodes) is therefore $2n-1$.
- Each tip node has a single branch, which connects it to its parent node.
- The root node, which is a special internal node that is older than all other nodes, has no parent branch, so it has only two branches. These connect it to its immediate descendant nodes (its children).
- All internal nodes in the tree other than the root have three branches—one connected to a parent node, and two connected to child nodes.

- Since every branch in the tree has a single child node, and each node in the tree except for the root has a parent branch, the total number of branches in the tree is $2n - 2$.

The function for the number of possible rooted phylogenies for n tips is:

$$f(n) = \frac{(2n-3)!}{2^{(n-2)}(n-2)!} \quad (2.1)$$

This gets big really, really fast as n increases. For a phylogeny of 5 tips there are 105 possible topologies. For a phylogeny of just 10 tips there are already 34,459,425 possible topologies. For a phylogeny of 50 tips there are 2.75×10^{76} possible topologies. This very large number of possible trees is a major challenge when it comes to inferring phylogenies from data—it is impossible to consider all possible topologies.

2.4 Rotating nodes

One of the most important things to keep in mind when interpreting a phylogeny is that the order of the tips doesn't convey any information (Baum and Smith, 2012). You can rotate any internal node, changing the order of the tips, and you still have the exact same phylogeny, just drawn a different way (Figure 2.7). The topology remains unchanged. It's the connections that tell us about the relationships. JD Laurence-Chasen, a former student in my invertebrate zoology course, made an excellent video about this—https://vimeo.com/148794860. The major implication of this is that you should never read a phylogeny across the tips to see, for example, which species are more closely related. You always need to look at the structure of the phylogeny itself.

2.5 The meaning of branch lengths

The meaning of branch lengths can vary from tree to tree (Figure 2.8). A phylogeny where branch lengths are scaled to time is known as a chronogram. Chronograms of organisms all sampled at the same time will be ultrametric, *i.e.*, the tip nodes will be flush. If tips are sampled at different times, for example when including fossils or sampling a rapidly evolving virus at different time points, they will not be ultrametric.

In many cases we don't have the information needed to scale branches to time, which requires calibration information such as fossils or the sampling dates of tips collected at different time points. Most published phylogenies therefore scale the branches according to the expected amount of evolutionary change in the characters under consideration. The longer the branch, the greater the expected change. The rate of evolution usually varies a bit across branches, so phylograms are not usually ultrametric.

Sometimes we are only interested in, or only have information about, the topology of the phylogeny. In this case we can draw the branches whatever length we want, and we call the phylogeny a cladogram. Cladograms can also be useful

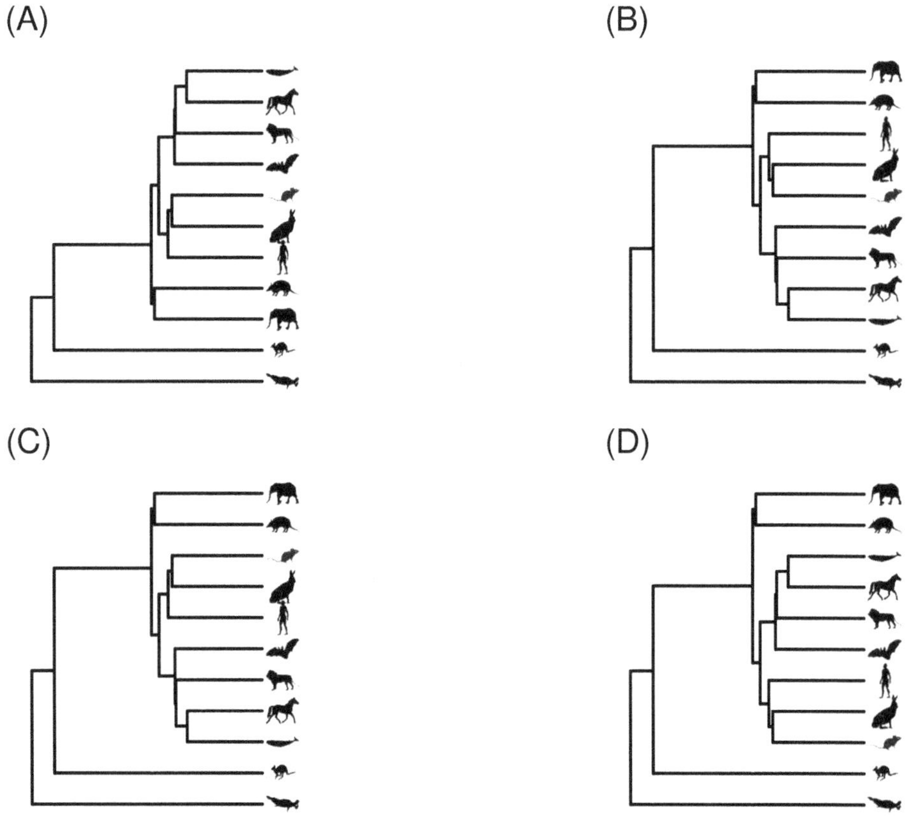

Figure 2.7: The exact same phylogeny, drawn a few times with different node rotations.

for showing annotations on branches in phylogenies with a wide variation in branch lengths, since there might not be enough room to write labels on very short branches.

It is always good practice whenever you show a phylogeny to indicate whether it is a chronogram, phylogram, or cladogram. Unfortunately, this is inconsistent in the literature, and not always clear from the figure legend or text. If you aren't sure what branch length means it is best to just ignore branch lengths and treat the phylogeny as if it were a cladogram.

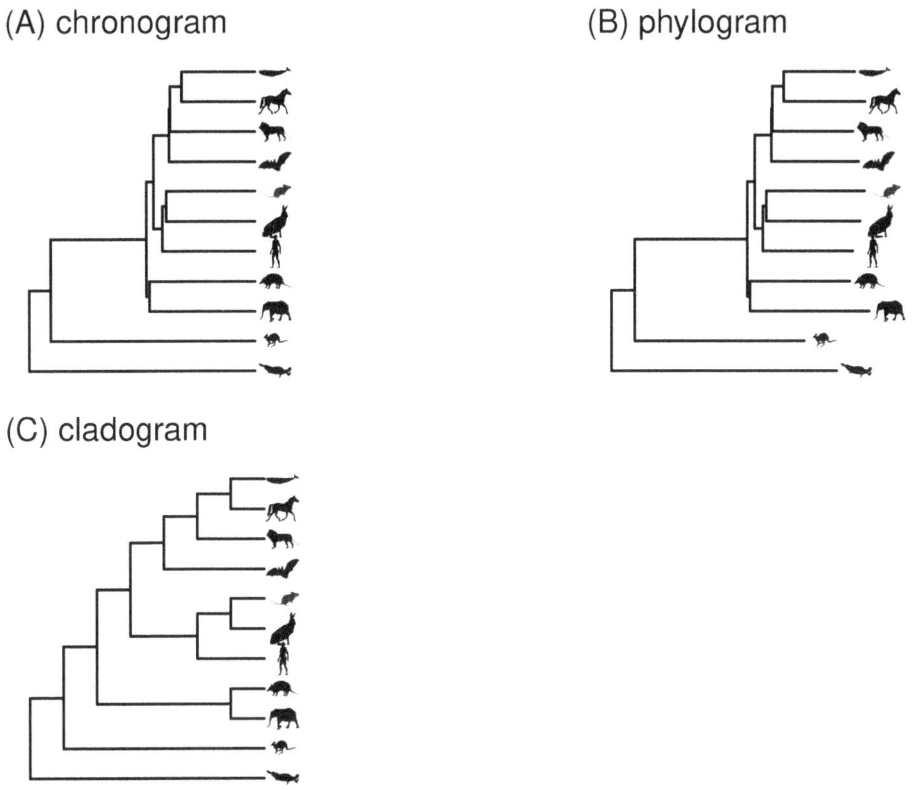

Figure 2.8: Several types of trees. In a chronogram, branch lengths are scaled according to time. In a phylogram, branch lengths are scaled according to expected amount of evolutionary change, which can differ across characters and branches. In a cladogram, the branch lengths have no meaning.

2.6 Naming nodes and groups

One of the most useful applications of phylogenies is to concisely refer to groups of organisms in light of their evolutionary history.

A convenient way to designate groups of organisms is by their most recent common ancestor (MRCA). One of the core concepts in phylogenetic biology is the clade—a group of tips that includes their MRCA and all of its descendants.

A group is said to be monophyletic (Figure 2.9A) if it satisfies these criteria (you don't need to use the double descriptor "monophyletic clade", since a clade is by definition monophyletic). Since internal nodes are included in this definition, you can think of a monophyletic group as a subtree that is formed by clipping the branch just below the MRCA.

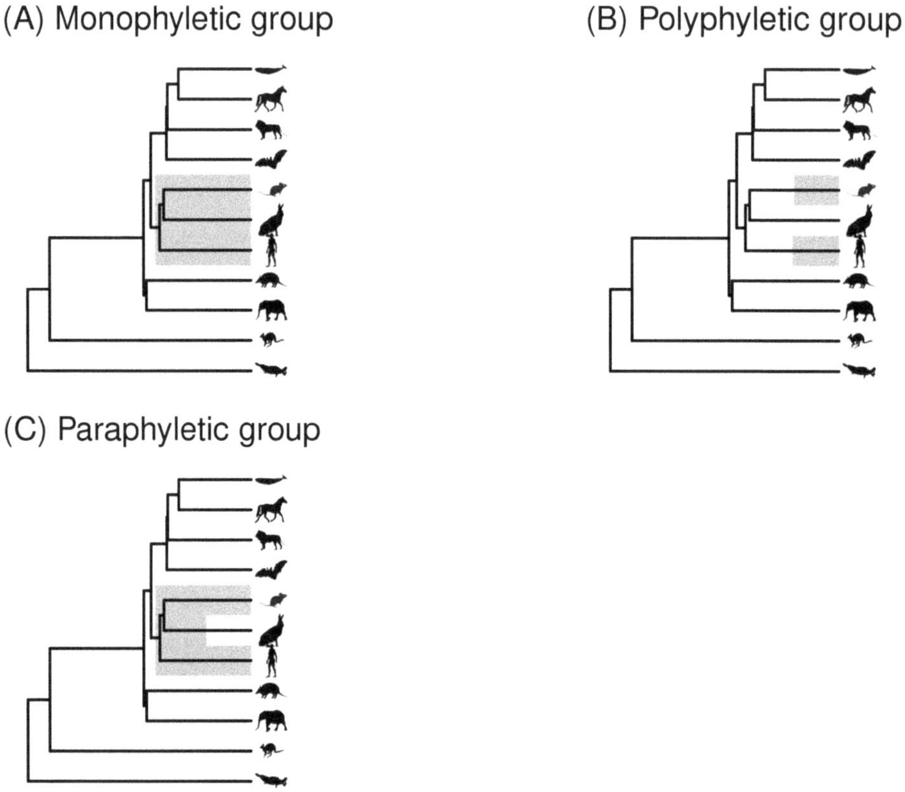

Figure 2.9: (A) A monophyletic group, i.e., a clade. (B) A polyphyletic group. (C) A paraphyletic group that does not include the hare.

Any internal node in a rooted phylogeny can be uniquely described as the MRCA of two or more tips. You can therefore use sets of tips to designate a particular internal node, and then use that node to define a clade. For example, we could clearly designate mammals as all the descendants of the MRCA of humans and platypus. This is a very compact and unambiguous way to name groups of organisms. In fact, an entire biological nomenclature has been built with phylogenetically defined names like these (de Queiroz, 2007; de Queiroz and Cantino, 2020).

Not all groups are monophyletic, and a group can fail to be monophyletic in two distinct ways. The difference between them is structural, and it turns on whether the group includes the most recent common ancestor (MRCA) of its members.

A group is paraphyletic if it includes the MRCA of its members but excludes some of that ancestor's descendants (Figure 2.9C). It is a clade with one or more subgroups removed. The group consisting of people and mice, together with the ancestral lineage they share, but excluding the hare that also descends from that ancestor, is paraphyletic: the common ancestor is included, but not all of its descendants are.

A group is polyphyletic if it excludes the MRCA of its members, combining two or more separate lineages (Figure 2.9B). Picking out people and mice as isolated lineages, without the shared ancestor and the branches that connect them, gives a polyphyletic group. Such groups often arise when organisms are grouped by a convergently evolved similarity rather than by common ancestry.

Note that this is not a matter of how one chooses to think about the same set of tips: the two cases are different groups, one that contains the ancestral lineage and one that does not. In practice the scale of the tree tends to determine which term is used. A large clade of 100 species that is intact except for a couple of species nested well within it that are left out is usually called paraphyletic, since it still contains the group's ancestor and merely drops a few descendants. A handful of tips scattered across a phylogeny of hundreds of species, whose deep common ancestor and the intervening lineages are excluded, is usually called polyphyletic.

2.7 Polytomies

Some phylogenies are not strictly bifurcating. An internal node with more than two branches connecting it to descendants is called a polytomy (Figure 2.10B-D). This can be due to uncertainty about branching order (a soft polytomy) or multiple divergence events in very quick succession, giving rise at effectively the same time to more than two lineages (a hard polytomy).

A phylogeny that consists entirely of a single polytomy is said to be entirely unresolved—it has no topological information. When laid out in a rectangular format, it looks like a comb (Figure 2.10D). When laid out in a slanted format it looks like a star. So you will hear fully unresolved trees referred to as comb or star phylogenies.

2.8 Rooting

All the phylogenies we have seen so far are rooted—we know what the oldest point is in the tree, and call it the root node. This special internal node has no branch connecting it to a parent node, and all other nodes in the phylogeny are its descendants. One consequence of having a rooted tree is that we know the direction of time along each branch—time proceeds from the root to the tips. This provides a clear parent-child relationship between nodes at the ends of each branch.

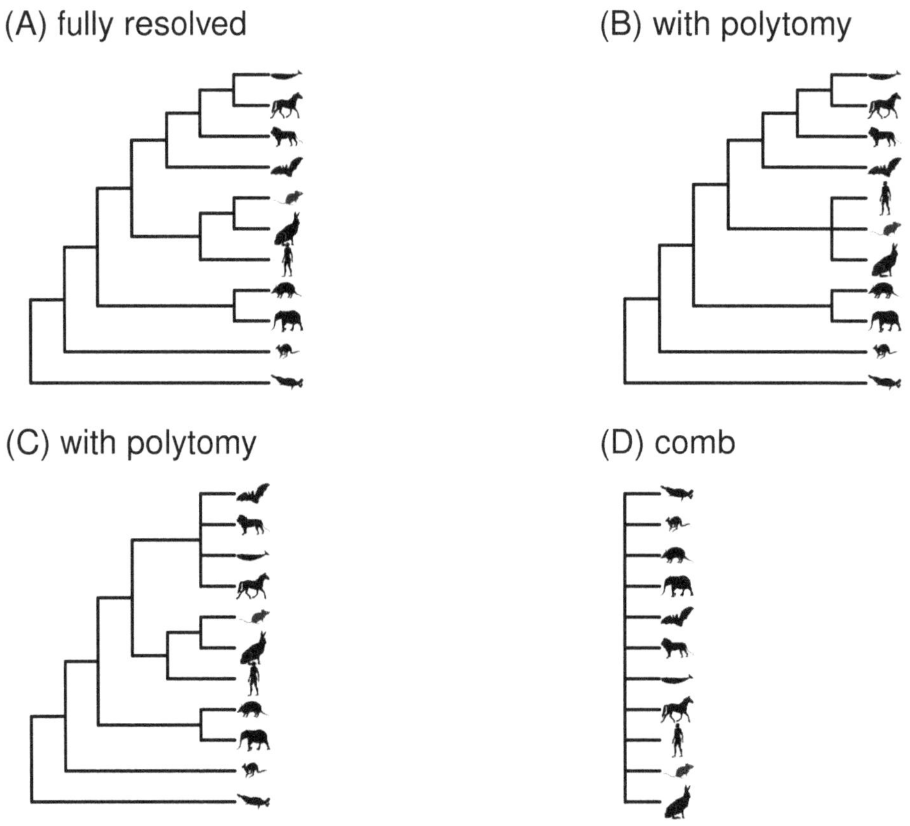

Figure 2.10: (A) The fully resolved mammal tree used in other figures. (B-C) Different polytomies created in this tree by collapsing some groups. (D) A fully unresolved comb tree.

Not all phylogenies are rooted. Sometimes we just don't care where the root is—some questions and methods don't depend on the root, so we don't have to bother placing it. Sometimes we just don't know—placing the root in a phylogeny takes information that sometimes we don't have. There are many phylogenetic studies focused on identifying the location of the root in various groups of organisms; this is often an interesting and important question.

Even when we don't know where the root of a phylogeny is, we often want to talk about other aspects of the tree, such as the topology and branch lengths. This means we need to consider how to work with unrooted phylogenies.

There are some basic things we can't take for granted in an unrooted phylogeny. We don't, for example, know which way time goes along the branches. We don't know which internal nodes are child nodes and which are parent nodes. We don't know which of the branches attached to an internal node connect to older nodes and which to younger nodes.

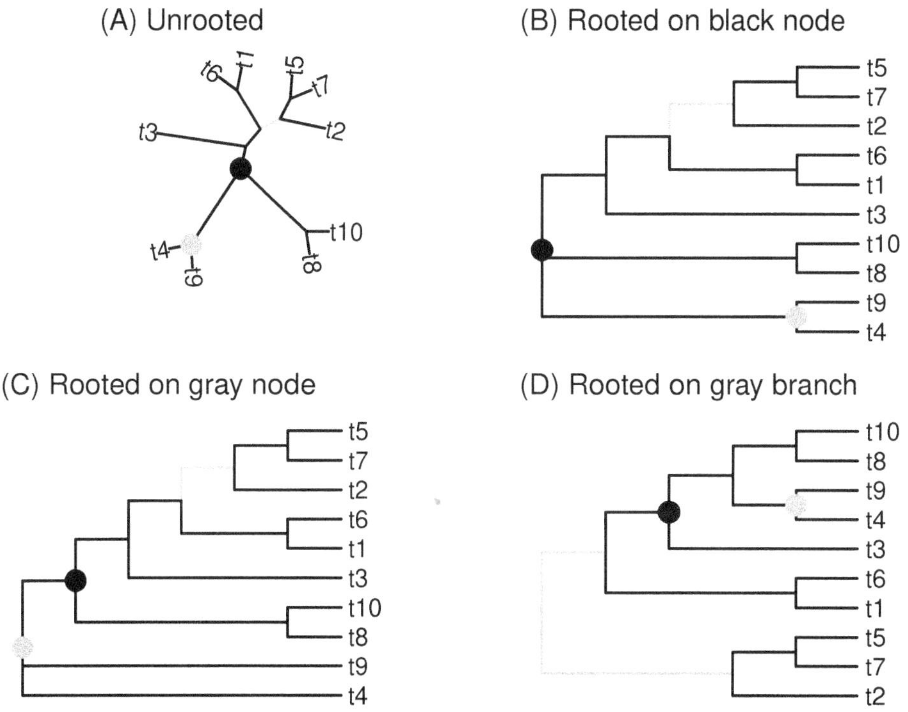

Figure 2.11: These four cladograms have the same tips and same topology; only the layout differs. (A) The first layout is unrooted. The other layouts are rooted on the black node (B), gray node (C), and gray branch (D). Each of these elements is in the exact same topological position in all trees. When the tree is rooted on the black or gray nodes, the base of the tree is a polytomy since these nodes have three branches attached to them. No nodes are added or removed when rooting on a node. When rooting on the gray branch, a new node is added along the branch. This new node is bifurcating.

The root could fall at any point in an unrooted phylogeny, either along a branch or right at a node. You can think of rooting a phylogeny as grabbing the point that you think is the root and dragging it until all the branches point away from it (Figure 2.11B-D). If you grab at some point along a branch (Figure 2.11D), this creates a new node that is the root. This new root node has two branches connecting it to descendants (these two branches arise by splitting the single branch along which the root was placed). Since it has no branch connecting it to a parent, it differs from other internal nodes in having only two branches connected to it (instead of three, or more in the case of polytomies). The end result of rooting along a branch is that you add a node to the phylogeny and the root is resolved (not a polytomy).

Things are a bit different if you root at an existing internal node (Figure 2.11B-C). Because the existing node becomes the root, the number of nodes in the tree remains unchanged. So far so good. But since internal nodes in a bifurcating tree all have three branches connected to them, when an internal node becomes the root the root is a polytomy. This seems like a pain, so why not always root along a branch? There are a few reasons. One is that rooting along a branch requires that we pick a specific point along the branch where the root goes, for example in the middle or somewhere else. That decision can take information that we don't have. Another reason is that the creation and destruction of nodes associated with rooting along branches gets cumbersome and problematic, especially when there are specific data associated with internal nodes.

What information can we use to root a phylogeny? There are a couple of approaches. One is to pick the midpoint—the middle of the longest path between two tips—as the root. If evolution proceeded at a perfectly uniform rate this could work well, but in practice rates of evolution are too variable across lineages for this to give a reliable result.

The more common approach to rooting a phylogeny is to use an outgroup (Figure 2.12). If the ingroup is the group of organisms you are interested in (*e.g.*, mammals), the outgroup is a set of organisms that you strongly believe do not fall within the ingroup. If you place the root anywhere in the outgroup, then the most recent common ancestor of the ingroup will be the root of the ingroup. Rooting this way works better when a larger number of outgroup species are considered, and the outgroup species include the organisms most closely related to the ingroup. Including too few outgroups is one of the most common mistakes in published phylogenies.

Always be careful interpreting root position when looking at a published phylogeny. Unrooted phylogenies are often drawn as if they are rooted, often by picking a random rooting point or making a good guess of where the root is. This is in part because rooted phylogeny figures tend to be easier to read than unrooted figures. If the figure legend or text doesn't state how the tree was rooted, it is best to assume that it wasn't rigorously rooted if this is critical to the analyses at hand.

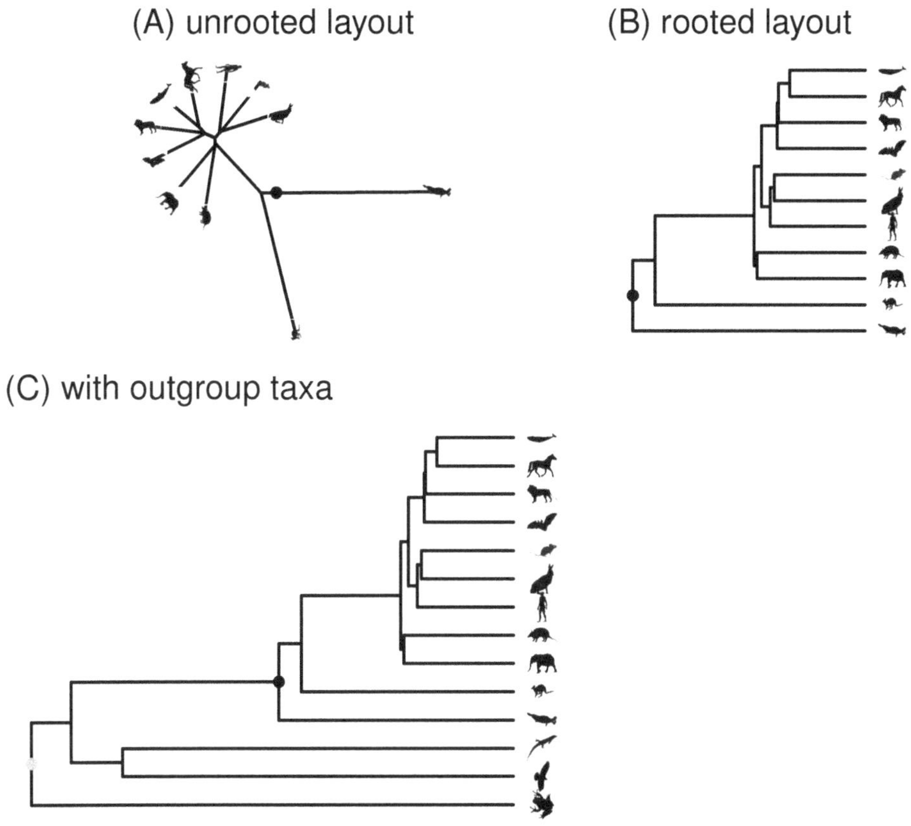

Figure 2.12: The root of the mammal tree is shown in black. (A) Unrooted layout. (B) Rooted layout. (C) Rooted layout, including outgroup. The root for the whole tree, shown in gray, is placed in the outgroup. The node where the ingroup is attached to the rest of the tree is the ingroup root. In this case, that is the black mammal root.

2.9 The information contained in phylogenies

Thinking about phylogenies as graphs gives us very explicit ways to think about what information can be contained in a phylogeny. In the fundamental graphical sense, a phylogeny is a pair of sets—a set of nodes and a set of branches—and annotations of those entities.

The topology of the phylogeny is the way in which nodes are connected by branches. If there are different connections in two phylogenies, then they have different topologies. If the annotations differ but the connections are the same, then the topologies are the same. Sometimes we know more and sometimes less about the topology. A fully resolved bifurcating phylogeny, where every internal node has two children, contains maximal information about the topology. A completely unresolved star phylogeny contains no information about topology.

Node annotations can include:

- Labels. On tip nodes these could be species names, gene names, or museum accession numbers for particular specimens. For internal nodes, they could be clade names (*e.g.*, Mammalia).
- Character states. The presence or absence of particular attributes (like hair), or the particular nucleotide at a particular spot in a particular gene (CGTA). Often the character states for tip nodes are observed from data and the states of internal nodes are estimated. There are exceptions, though. One can, for example, estimate the states of tips for which no observations are available.
- Geographical locations.
- Whether the node is the root.

Branch annotations can include:

- Length. The value and unit of length can differ, for example it can be time (as in a chronogram) or expected amount of change in a particular set of characters (as in a phylogram).
- Directionality. An indication of which direction time goes in along each branch. This only applies in the context of rooted phylogenies.
- Events. These could include a character change (such as the gain of a placenta) or a geographical change (such as an intercontinental dispersal event). Nodes at each end of a branch need not be different in the relevant trait to have change events along the branch. For example, at a particular gene site a parent and a child node could both have an A. There could be two or more events along the branch that are compatible with this pattern, for example a change from A to C and then a change from C back to A.
- Branch frequencies. This is how topology support values, such as bootstraps and posterior probabilities (which we'll discuss later), are stored.

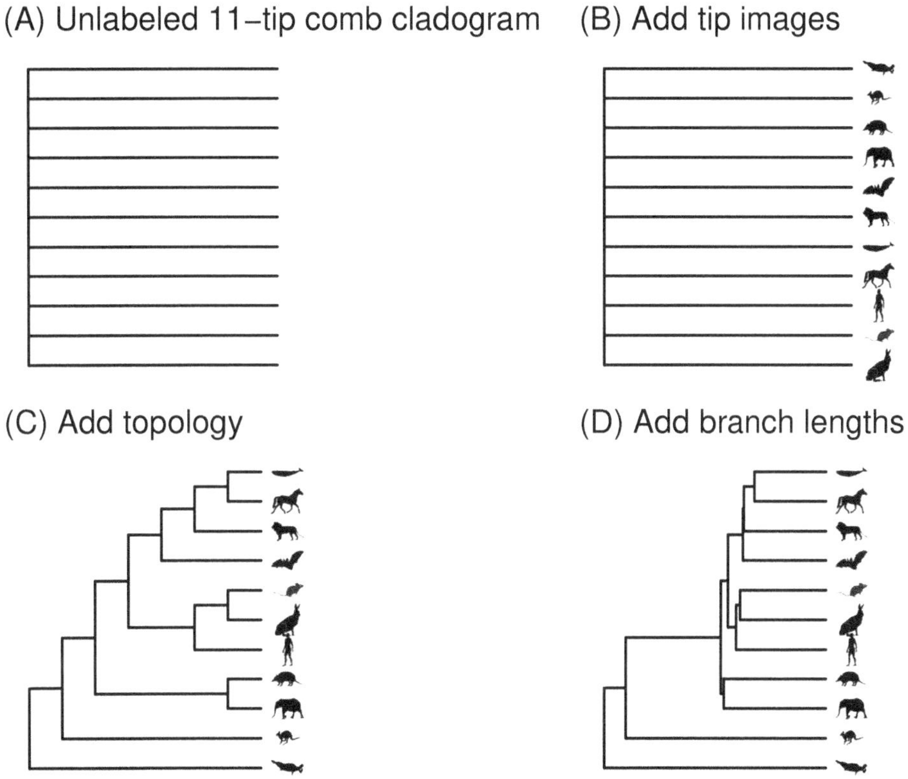

Figure 2.13: Adding information to a phylogeny makes it more specific.

Not all phylogenies have all this information. The minimum possible information a phylogeny could contain is just the number of tips. With this, you could draw an unresolved, unlabeled cladogram (Figure 2.13A). You could then start layering information onto that phylogeny. For example, you could next add images denoting the species at the tips (Figure 2.13B). That tells you what species are in the phylogeny, but nothing about how they are related. Next you could add internal nodes and branches that indicate the topology of the phylogeny (Figure 2.13C). Once you have topology, you could display branch attributes. For example, you could scale the branch lengths by time (Figure 2.13D).

At each step in this process of layering on information, the phylogeny is compatible with a very large set of possible phylogenies. At one extreme, Figure 2.13A is compatible with any phylogeny with any branch lengths for any 11 species or organisms or genes. Each added piece of information narrows that subset. When we label the tips as in Figure 2.13B, it is compatible with any phylogeny with any branch lengths for those specific species. And so on. More information provides more specificity. We could go well beyond Figure 2.13D, for example by labeling internal nodes or showing character states.

The amount of information in a phylogeny varies greatly depending on a variety of factors, including what information is available, what the question at hand is, and what makes the most sense for the focused description of the biology at hand. One investigator may go to great lengths to calibrate branch lengths, for example, while another couldn't care less about branch lengths and is only interested in topology.

2.10 Representation

So far we have focused on phylogenies in the abstract sense as mathematical graphs and their associated annotations, and in the concrete sense as tree-like drawings that represent these relationships. To work with trees of any significant size, we need to also represent them computationally. How can you store and manipulate a whole tree and all its associated annotations?

One way is to store a text representation of the tree. This is convenient because text files are easy to work with, and can be viewed and edited with a variety of existing text editors. Newick, a phylogenetic data format ironed out over dinner at Newick's Lobster House, does exactly this. The Newick format is widely used for storing phylogenies in text files, and is supported as a way to read and write trees by almost every phylogenetic software tool. The basic idea is to designate each clade within a pair of parentheses.

```
# Define the newick text that includes the tip labels and
# tree topology
newick_text = "(((A,B),(C,D)),E);"

# Create an ape phylo object from the tree text
phylo_tree = read.tree( text=newick_text )
```

```
# Plot the newick_tree phylo object using the ggtree library
# Draw the tree
ggtree(phylo_tree ) +
  # Add the node numbers
  geom_text2(aes(label=node), col="gray", nudge_x=0.1 ) +
  # Add the tip labels
  geom_tiplab( aes(label=label), offset=0.3, col="black") +
  # Add points on nodes
  geom_nodepoint(col="gray") +
  geom_tippoint(col="gray")
```

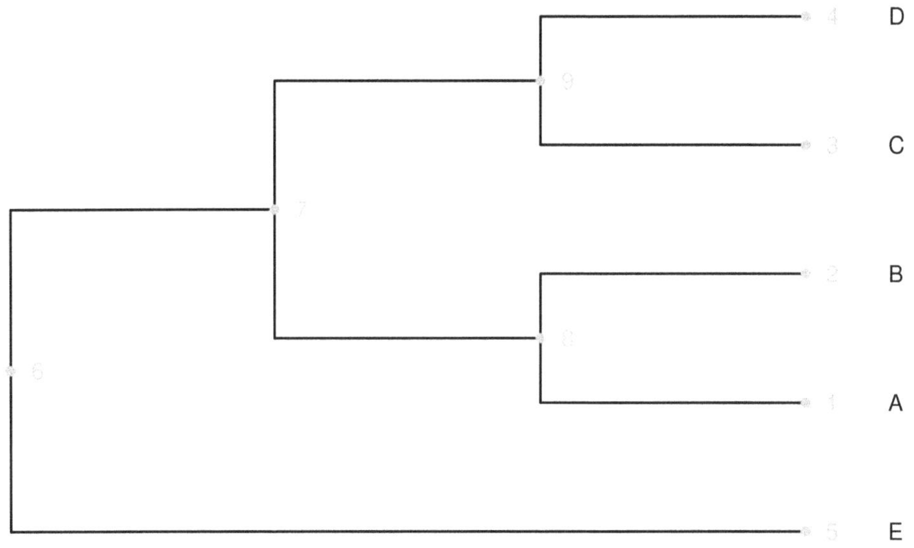

Figure 2.14: Defining and drawing a Newick tree. Tip labels (letters) are black, nodes and node numbers are gray, and branches are black.

In the R code above, for Figure 2.14, you can see how to define a Newick tree, convert it to a different type of representation, and then draw that. Most of the figures in this text were made using similar code.

Considering just the Newick specification of the tree, `"(((A,B),(C,D)),E);"`, you can see it built up as a series of clades. The inner-most sets of parentheses define two clades, `(A,B)` and `(C,D)`. The next set of parentheses out indicates that these two clades together form a larger clade, `((A,B),(C,D))`. Finally, the outermost parentheses indicate that the clade `((A,B),(C,D))` is sister to `E`. Commas separate sisters within clades, and the whole thing is finished off with a semicolon. The format also allows for labels of internal nodes, and the specification of branch lengths. Joe Felsenstein wrote an interesting description and history of the Newick format, which includes details on how to store other types of information in the file format—https://evolution.genetics.washington.edu/phylip/newicktree.html.

As versatile and simple as Newick is for storing trees in files, it isn't great for storing trees in computer memory where you want to do things with them. To build and analyze trees it is better to have a format that has a more direct representation of nodes, branches, and their annotations. This allows us to directly encode the information noted in the section The information contained in phylogenies, and to readily extend the data objects as needed.

The most widely used format for storing phylogenies in the R programming language is as a `phylo` object from the excellent `ape` library (Paradis et al., 2024). The `read.tree()` function in the code block above creates a `phylo` object called `phylo_tree` based on the phylogeny we specified as text and named `newick_text`. In a `phylo` object, each node of a phylogeny has a unique number. The first consecutive node numbers, from 1 to n where n is the number of tip nodes, are allocated to the tip nodes. The internal nodes are numbered consecutively from there, which in a bifurcating tree will be nodes $n+1$ to $2n-1$. The assignment of the node numbers within these ranges is arbitrary, and there is no guarantee that the same nodes will have the same numbers each time the tree is read. The numbers of the nodes that were given when we created the `phylo` object from the Newick text are shown in gray in Figure 2.14.

Below we take a quick peek inside the `phylo` object we created above. The intent isn't to learn how to manipulate trees in R quite yet, but to just show you how trees can be stored in computer memory. First, let's take a look at the structure of the `phylo` object to see what variables it contains within it:

```
str( phylo_tree )
```

```
## List of 3
##  $ edge     : int [1:8, 1:2] 6 7 8 8 7 9 9 6 7 8 ...
##  $ Nnode    : int 4
##  $ tip.label: chr [1:5] "A" "B" "C" "D" ...
##  - attr(*, "class")= chr "phylo"
##  - attr(*, "order")= chr "cladewise"
```

We can see that there are three slots (each designated with a `$`) within this `phylo` object. `Nnode` is just the number of internal nodes. The real magic is `edge`, which encodes the topology of the tree. It is just an array with one row per edge (*i.e.*, branch), and two columns. The value in column 1 is the number of the node that is the parent of the branch, and the value in column 2 is the number of the node that is the child of the branch. Here are the complete contents of the `edge` slot:

```
phylo_tree$edge
```

```
##      [,1] [,2]
## [1,]    6    7
## [2,]    7    8
## [3,]    8    1
## [4,]    8    2
## [5,]    7    9
## [6,]    9    3
## [7,]    9    4
```

```
## [8,]    6    5
```

So, in this case edge number 1 (the first row) connects node 6 to node 7. Indeed, we can see just such an edge in Figure 2.14. Try to identify each of the other branches in the array in the tree figure.

You can discern all sorts of things about the tree from this simple array. The tips are nodes that are in column 2 but not in column 1 (since they aren't parents of any other nodes). The root is the node that is in column 1, but not in column 2 (because it doesn't have a parent). If a node occurs in column 1 more than twice, then it is a polytomy (because it has more than two branches connecting to child nodes). Note that this data structure is intrinsically rooted—there is a parent-child relationship set up between the nodes at each end of each edge. But you can still use it to store unrooted trees by picking an arbitrary internal node as the root, and just keeping track of the fact that you don't know where the root is.

There is one other important piece of information that we specified in the Newick file—the tip labels. Those are located in another slot within the `phylo` object:

```
phylo_tree$tip.label
```

```
## [1] "A" "B" "C" "D" "E"
```

We can see that this is just a vector of labels. They correspond to the labels of nodes one through five, all the tip labels, in consecutive order.

That's it. That's all the information we specified in the Newick file, all the information in the `phylo` object, and all the information that is needed to draw Figure 2.14. The Newick tree is easy to store and manipulate with simple text tools, the `phylo` object is a powerful and flexible way to represent the tree in computer memory, and the figure is easy to read at a glance.

2.11 Additional reading

I used `ggtree` (Yu et al., 2026) to draw the trees in this chapter. If you are interested in learning more about these tools, see:

- The ggtree book (Yu, 2022)
- ggtree vignettes—https://bioconductor.statistik.tu-dortmund.de/packages/3.1/bioc/vignettes/ggtree/inst/doc/ggtree.html

Chapter 3

Simulation

Here we will build the machinery to implement models of DNA evolution. Our first application of these models will be to simulate data.

3.1 Models

Models are representations of processes. They are abstractions in the sense that they leave out things that aren't thought to be important, and they are idealized in the sense that they are deliberately simplified (Godfrey-Smith, 2014, 21).

For some applications, it doesn't matter if model structure reflects actual underlying processes, as long as it generates useful output (Breiman, 2001). For example, if data scientists at a large retail chain are trying to predict how much toothpaste they need to stock at each store, they likely don't care if their models properly consider purchasing rates and all the other things that impact stock, so long as the model does a reasonable job of making useful predictions. In science, though, we often care very much about the model because many of our questions have to do with the mechanisms that underlie the processes we are modeling. We don't just want models that give us the right answer, we often want models that give us the right answer for the right reasons.

Statisticians often note that "All models are wrong but some are useful", a common aphorism expanded from a quote by Box (1976). The goal of a model isn't to be "right" in the sense of being a perfect explanation of data. Real data are far too complex for any model to do this. Instead, the most useful models strike a trade-off between simplicity and adequacy. They are as simple as possible, while adequately describing the phenomenon of interest. Adequacy is often in the eye of the beholder—one scientist will be perfectly happy with a model that makes reasonable rough approximations of a system, while another scientist may be interested in second- and third-order effects for which more complexity is needed for adequate explanation.

Cartography is an interesting analogy for the way we will use statistical models. A "perfect" map would basically be a copy of the whole world, which wouldn't be that much more useful than the world itself already is for many of the things you would like to do with a map. So all maps are simplifications (Figure 3.1). The simplification is often what makes the map useful.

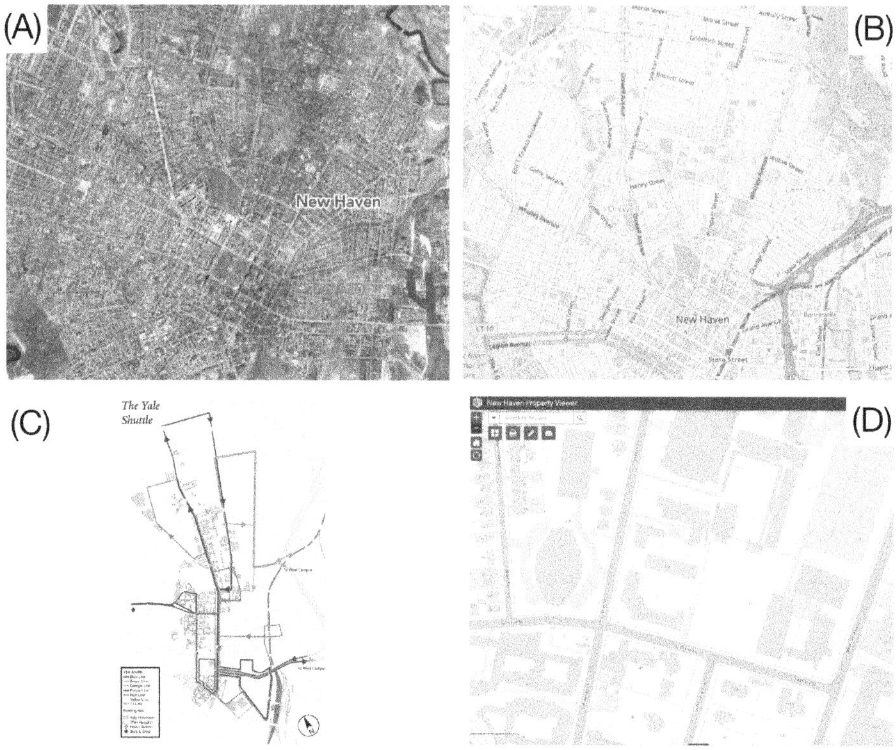

Figure 3.1: Four maps of the Yale campus, varying in complexity and focus. (A) Aerial image of New Haven, including much of Yale campus, from 1934. This image has a very large amount of information. (B) A street map of the same region; map data © OpenStreetMap contributors, available under the Open Database License. It has less information, but is more useful for some tasks such as navigation. (C) An even more simplified map, focused on showing the Yale Shuttle routes. (D) The New Haven property map of the region around Osborn Memorial Laboratory, showing property lines and plot numbers. Like (C) it is simple, but reflects different decisions about which information to discard or retain. This figure is inspired by the London maps that David Swofford uses in his own talks to make the same points.

Let's examine one of the most common models, the linear model:

$$y = mx + b \tag{3.1}$$

Here, y and x are variables. The model posits a linear relationship between x and y. If you plot their correspondence in a plane you will get a line (Figure 3.2). m and b are model parameters. m is the slope of the line. It captures how much change there is in y for each unit of change in x. b is the intercept. It is the value of y when $x = 0$.

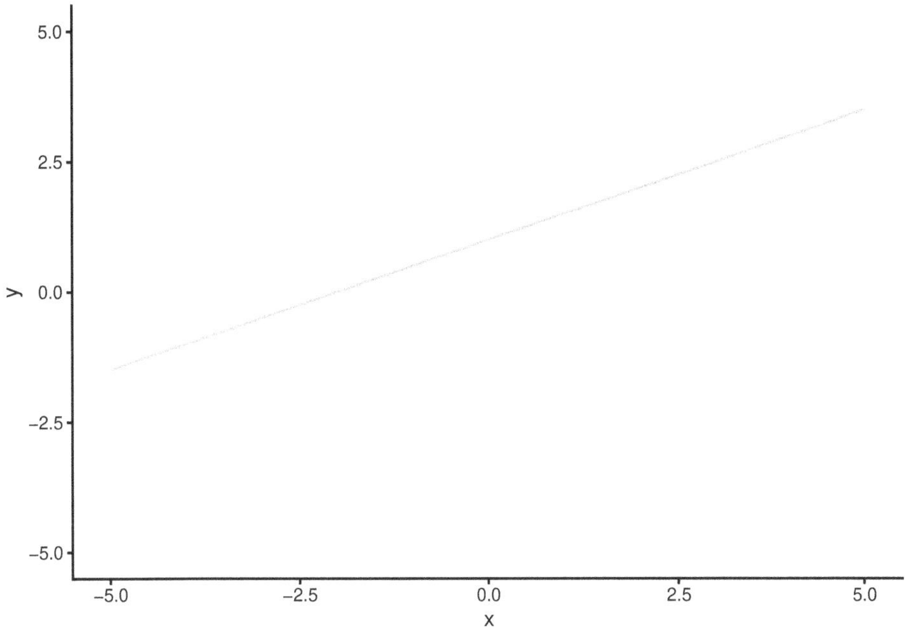

Figure 3.2: A linear model with $m = 0.5$ and $b = 1$.

There are a variety of useful things we could do based on these relationships between the model, model parameters, and values. Let's consider x to be a variable that tells us something about the past and y to be a variable that tells us something about the present or future. Use cases then include:

- If we clamp the model (linear) and model parameters (specific values of m and b) according to prior knowledge, and clamp y according to observed data, we can estimate x. In this case the model is like a time machine that allows us to look into the past.
- If we clamp the model (linear) and variables (x and y) according to prior knowledge, we can estimate model parameters m and b. In this case the model is like an instrument that allows us to peer inside processes based on their inputs and outputs.
- If we clamp the model (linear) and model parameters (specific values of m and b) according to prior knowledge, and clamp x according to observed data, we can estimate y. In this case the model is like an oracle that predicts the future.

- If we clamp the model (linear) and model parameters (specific values of m and b) according to prior knowledge, and clamp x according to values we make up, we can simulate y. In this case the model is like a world-builder that tells us what we would expect to see under the specified conditions. This is very helpful for getting a sense of whether our models are working well (Do they generate results that look like what we see?), examining counterfactual situations that don't exist, or building datasets to test software tools.

There are some deep connections here. For example, predicting the future is basically just simulating the future based on our model and what we already know about the past and present.

The models that we will use in phylogenetic biology tend to be more complex than the linear model, but this general perspective of clamping and estimating different things still holds.

3.2 A simple model of evolution

Let's start with a simple model of DNA evolution. At first we will consider only a single nucleotide position along a single branch in a phylogeny (Figure 3.3). The goal is to build an intuitive integrated understanding of the mathematical and statistical relationships among:

- Model structure
- Model parameters
- Branch length
- State at the start of the branch (the nucleotide at the parent node)
- State at the end of the branch (the nucleotide at the child node)

When DNA is replicated, the appropriate nucleotide is usually incorporated. Some fraction of the time, at rate μ, an event occurs where the appropriate nucleotide is replaced with a random nucleotide instead. In our model, the probability of selecting any of the nucleotides during one of these random replacement events is uniform (picking a C is just as probable as picking a G, for example), and the new nucleotide doesn't depend in any way on what nucleotide was there before. It is as if you had a bag containing a large number of C, G, T, and A nucleotides at equal frequencies. As you built the new DNA strand, every so often you would replace the nucleotide you should be adding with one you instead select by reaching into the bag and picking at random.

Not all replacement events will result in an apparent change. Sometimes the appropriate nucleotide is selected by chance, even though it was picked at random. If, for example, the appropriate nucleotide was an A, under this model 1/4 of the time that a replacement event occurs, an A is selected by chance and there is no apparent change. In such a case, there has not been a substitution, just a replacement in kind. If the A is replaced with any of the other three nucleotides we say there has been a substitution. Because three of the four possible outcomes of a replacement event result in a substitution, the substitution rate is $(3/4)\mu$. Because some events result in no apparent change, substitutions are

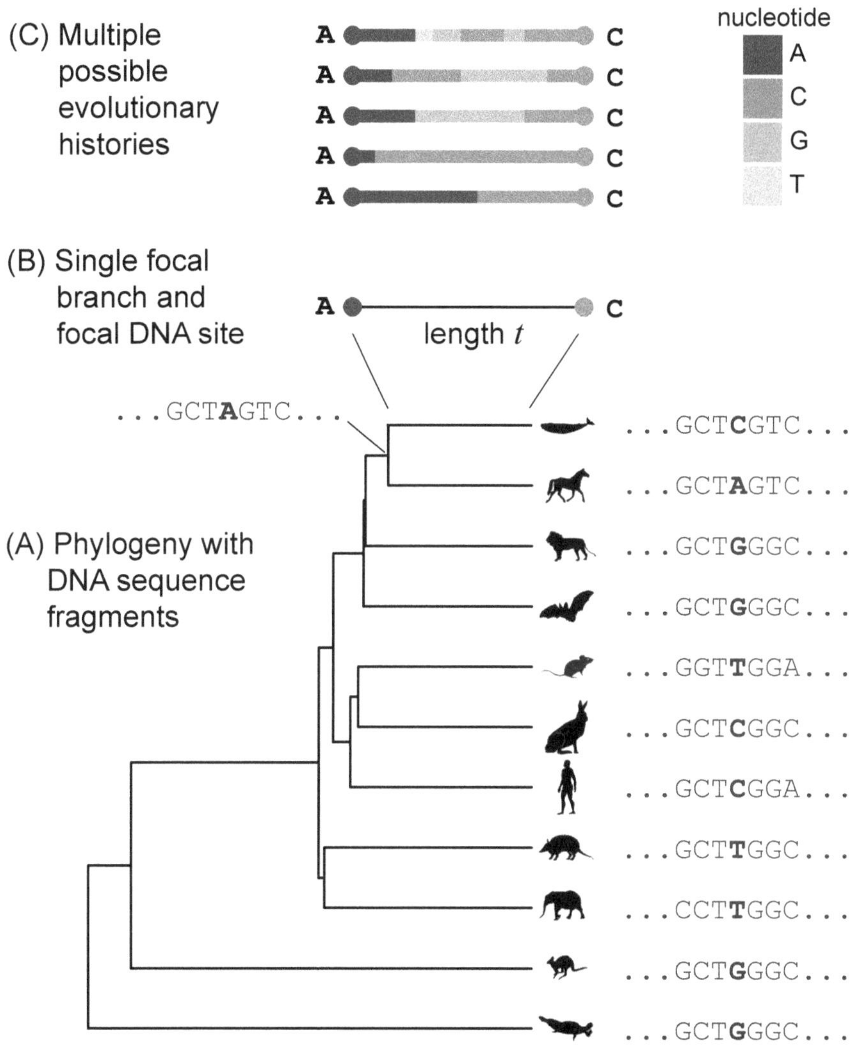

Figure 3.3: Our current goal is to model the evolution of a single site in a DNA sequence along a single branch in a phylogeny. (A) An example phylogeny, with DNA sequence fragments shown at the tips and one internal node. The site under examination is shown in bold, and the branch under examination (at the top) is thicker than the rest. (B) A closeup of the focal branch, and the state of the focal site at its ends (the parent and child nodes). (C) Multiple mutational histories that are consistent with the starting and end states shown in (B), i.e., a change from A to C.

only a subset of events and the substitution rate is *lower* than the replacement rate.

It might seem a bit odd to consider replacement events that don't result in substitutions, but this follows naturally from a central feature we specified for the model—the new nucleotide doesn't depend in any way on what nucleotide was there before. If we had a process where replacements always resulted in substitutions, then excluding the replacement-in-kind events would require knowing which nucleotide should be placed so that we *don't* select it.

3.2.1 Expected amount of change

For the simple process described here, there are two things to consider if we want to know the amount of evolutionary change. The first is the rate μ, and the second is the time over which the evolutionary process acts. In our example here, that time is the length of the branch under consideration in the phylogeny.

In Figure 3.4 each horizontal bar is a simulation over the same time interval (0-100 time units). Each black line on the bar is a replacement event randomly introduced by the computer according to our model. We use a different value of μ for each simulation (as indicated on the vertical axis). In the bottom bar, where $\mu = 0$, there are no replacements (black lines) and therefore no substitutions (the whole bar stays a single shade, corresponding to the starting nucleotide A). There are more replacement events as μ increases along the vertical axis.

As μ increases (going up on the vertical axis), the number of replacement events over the same time interval increases (Figure 3.5). This reflects the simple linear relationship $n = \mu t$, where n is the number of expected replacement events.

Because of the linear relationship between the number of replacements and the product μt, rate (μ) and time (t) are conflated. In many scenarios you can't estimate them independently. If there are a small number of replacements, for example, you can't tell if there is a low rate over a long time interval, or a high rate over a short interval. Both would give the same resulting number of changes n.

Because rate (μ) and time (t) are so often conflated in phylogenetic questions, branch lengths are usually rescaled and given in units of the expected number of substitutions per site rather than absolute time (years, months, *etc.*). The factor relating μt to the expected number of substitutions depends on the model, because not every replacement is a substitution. Under the simple model considered here, the substitution rate is $(3/4)\mu$, so a branch of duration t carries $(3/4)\mu t$ expected substitutions per site. You will often see the resulting branch lengths as the scale bar of published phylogenies (Figure 3.6) (Zapata et al., 2015).

The exception is when you have external information, such as dated fossils, that allow you to independently estimate rates and branch lengths in terms of actual time. Sometimes deconfounding μt isn't important to the primary question of

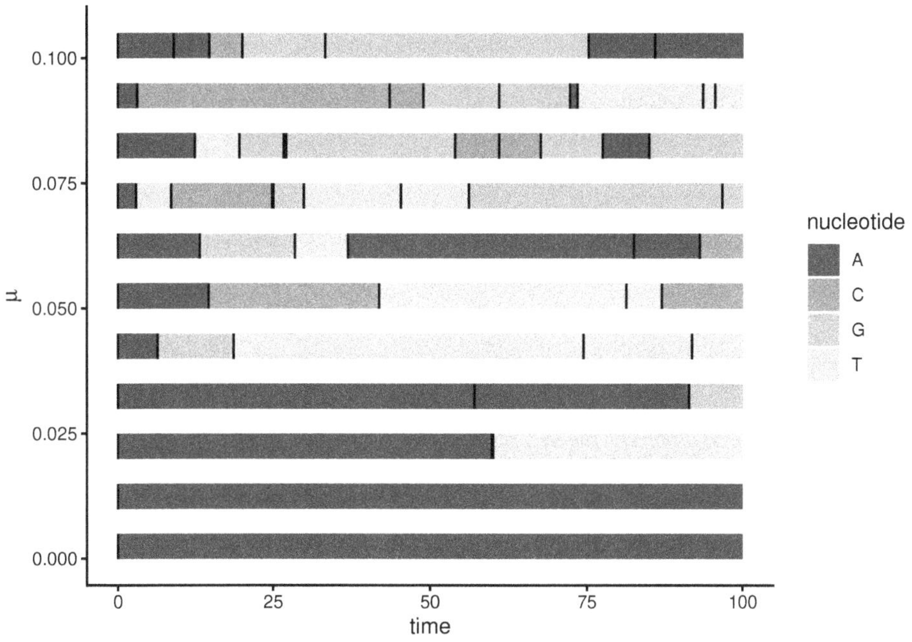

Figure 3.4: Each horizontal bar is a simulation of evolution of a single nucleotide position through time, t, for a specified value of μ. Each simulation starts out as an A. Black vertical lines correspond to replacement events, which don't all lead to substitutions (a shift to a different shade, marking a new nucleotide state).

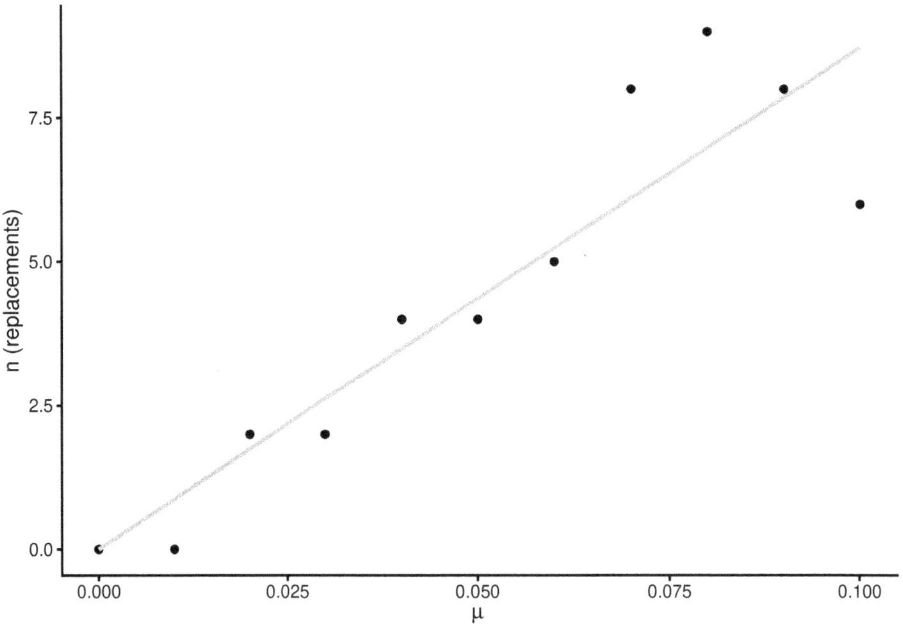

Figure 3.5: The number of replacement events increases linearly with the replacement rate μ. This plot is from the same simulation as that shown in Figure 3.4. The line is a linear model fit to the data. Since $n = \mu t$, and in this case $t = 100$, the slope of n on μ is estimated to be near 100.

the investigator, sometimes it would be nice to know but can't be done, and other times (such as in papers that date trees) it *is* the central question.

3.2.2 Expected end state

The machinery above shows how a model can clarify the way we think about the expected amount of change along a branch. Often, though, we want to know what the probability of a given end state is given a starting state, a model, and the amount of time elapsed. One way to anchor this question is to think about the extremes—what do we expect after a very small amount of change (either a short time or a slow rate of change, or both), and what do we expect after a large amount of change?

The situation is most clear after a small amount of change (when μt is small)—we expect the end state to be the same as the starting state. If we start with an A, for example, if there is very little change we expect to end with an A (Figure 3.7, left side). In this situation, the starting state tells us a lot about the end state. Not much else matters.

What should we expect, though, if there has been a large amount of change (when μt is large)? Can we know anything at all? It turns out that we can. If there have been many replacements, one after the other, then the initial starting state doesn't matter because whatever was there initially will probably have been replaced multiple times. It is as if it had been erased and written over. If the starting state doesn't contain information about the end state, what does?

Since replacements are coming from the bag that you are picking the nucleotides at random from, that bag has information about the expected states after a large number of changes. Given enough evolutionary time, our simple model will lead the expected frequency of each nucleotide in the evolving sequence to be the same as their frequencies in the bag. Since we specified that you have the same chance of grabbing any nucleotide from the bag, eventually the probability of having each of the four nucleotides is the same, 25% (Figure 3.7, right side). If you started with a sequence that had an A and let it evolve 100 times, after enough evolutionary time had passed to reach equilibrium you would expect to get 25 C's, 25 G's, 25 T's, and 25 A's.

3.2.3 Analytical approach

In all the examples above, I simulated replacement events by randomly generating them at the specified rate. To get the end state, I then just retained the last state. You could take this approach to assessing the probability of different outcomes in actual analyses, but it gets very computationally expensive. It would be better to have an equation to solve for the probabilities directly. That isn't always possible for a model, but it is in this case.

The process of change that our simple model describes is similar to compound interest. We have something in hand, apply a process to it, then take the output and apply that same process again. Over and over. In the case of

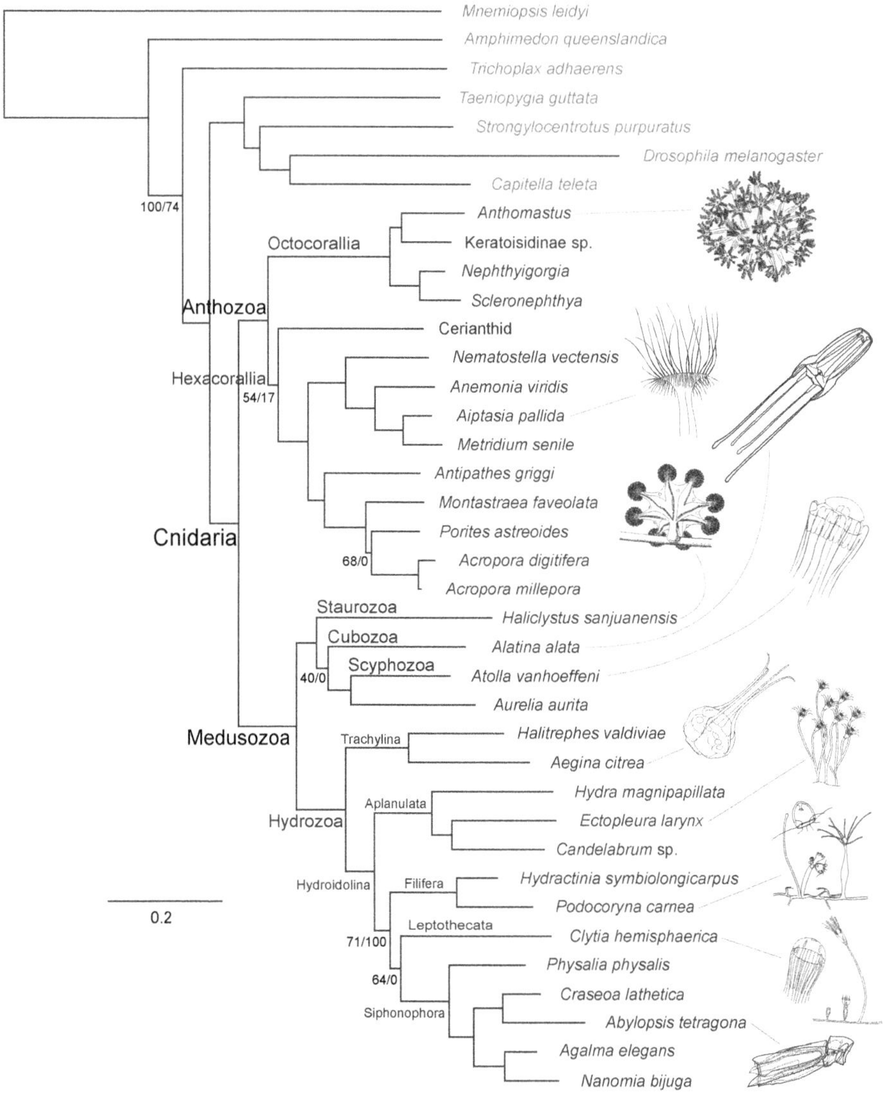

Figure 3.6: A published phylogeny (Zapata et al., 2015) with a scale bar indicating branch length in terms of the expected amount of evolutionary change, rather than absolute time.

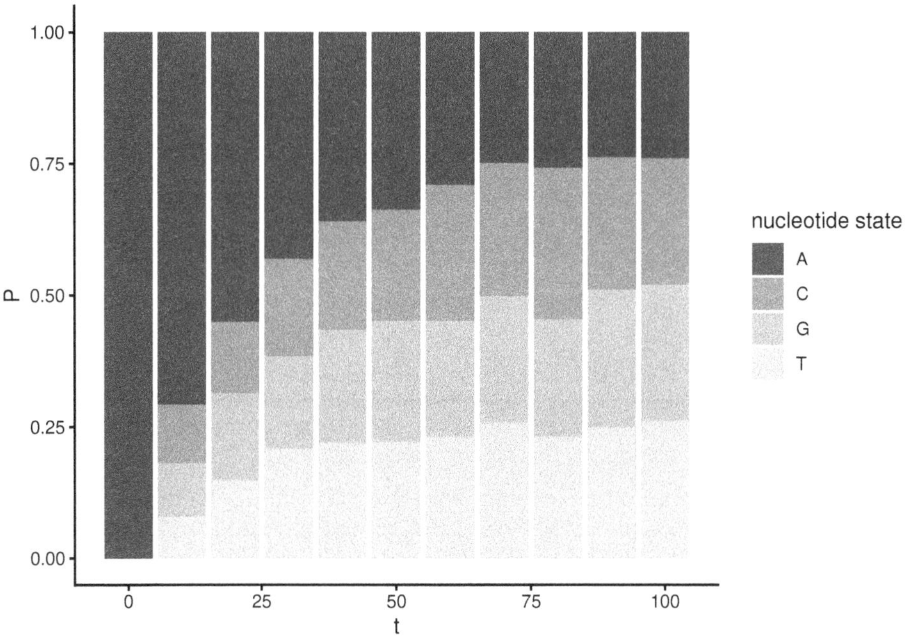

Figure 3.7: Stacked bar plots indicating the frequency of each nucleotide after simulated evolution for a specified amount of time. The rate of evolution is $\mu = 0.050$. There are 1000 replicate simulations for each value of time. At time $t = 0$ (no evolution), the end result is always the same as the initial value, which is fixed at A in these simulations. As the length of time increases, the four nucleotides converge on equal proportions of 0.25 each.

compound interest, that something is money and the process is growth. In the case of our model, the something is a DNA site and the process is mutation. In both cases, we take as inputs a starting state, a rate of change, and an amount of time, and as an output get the expected end state. To get the expected end state as a function of time elapsed, given a rate of change, we can use exponential functions. For example, here is the exponential function for calculating a balance at time t, written $B(t)$, given the initial balance B_0 and an interest rate r:

$$B(t) = B_0 e^{rt} \tag{3.2}$$

For our sequence evolution model, we need two exponential functions (Swofford et al., 1996):

$$P_{\text{same}}(t) = \frac{1}{4} + \frac{3}{4}e^{-\mu t} \tag{3.3}$$

$$P_{\text{different}}(t) = \frac{1}{4} - \frac{1}{4}e^{-\mu t} \tag{3.4}$$

Equation (3.3), which we write as $P_{\text{same}}(t)$, shows the probability of the final state being the same as the beginning state. So if you start with an A, this would give you the probability of remaining an A after time t given a specific value of μ. Equation (3.4), which we write as $P_{\text{different}}(t)$, is the probability of each of the three end states that are different from the starting state. If you start as an A, this is the probability of changing to a G, for example. It is also the probability of C to T, G to A, *etc.* Note that these are the probabilities of *each* of the four possible end states, so they sum to one across the four states as $P_{\text{same}}(t) + 3P_{\text{different}}(t) = 1$ for any t.

Consider what happens to these equations in the extremes we considered above when examining Figure 3.7. If μ or t are zero, we expect no change (Figure 3.8, left side). In that case we get e^0, which is 1. $P_{\text{same}}(t)$ (Equation (3.3)) becomes $1/4 + 3/4$, which is 1. So there is a probability of 1 that, after no change, the end state is the same as the beginning state. Likewise, $P_{\text{different}}(t)$ (Equation (3.4)) becomes $1/4 - 1/4$, which is 0. So after no change the end states that differ from the beginning state each have probability 0.

Now consider the case after infinite change (or just a large amount of change, as in the right side of Figure 3.8). If μ or t are infinity, then $e^{-\mu t}$ becomes $e^{-\infty}$, which is 0. In that case, $P_{\text{same}}(t)$ becomes $1/4 + 0$, which is simply $1/4$. Likewise, $P_{\text{different}}(t)$ becomes $1/4 - 0$, which is also $1/4$. So all the nucleotides (the one that you started with, and the three other states that substitution can lead to) all have the same equal frequency of $1/4$. This reflects the fact that the frequency of drawing each of these from the bag was $1/4$.

We can reorganize things a bit (Figure 3.8) to get a plot like that of Figure 3.7, but derived from Equations (3.3) and (3.4) instead of from simulations of changes along branches (Figure 3.9).

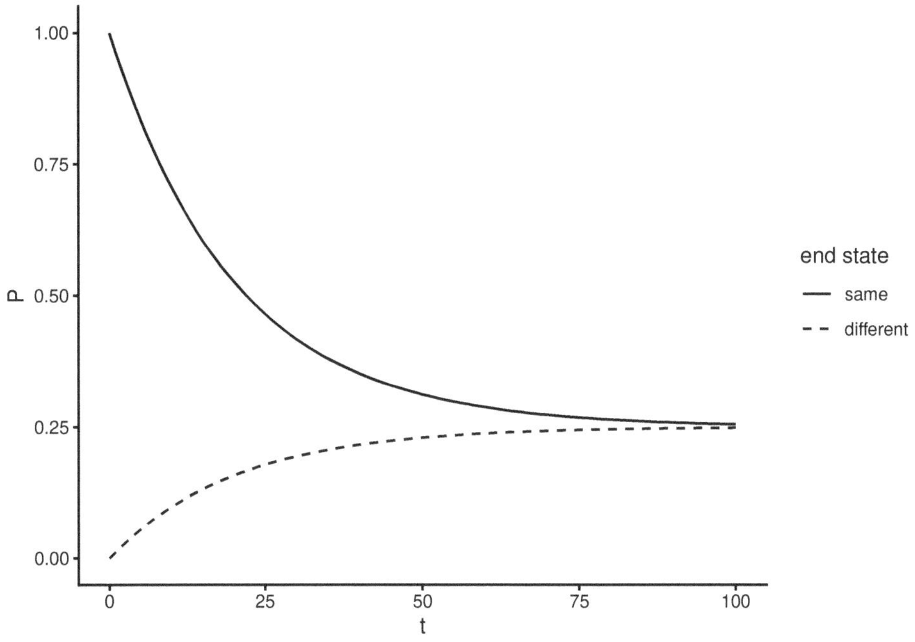

Figure 3.8: The probability of observing a particular end state at time t, given the start state A and $\mu = 0.05$. The solid line, labeled same, is $P_{\text{same}}(t)$, the probability of observing the original start state (Equation (3.3)). The dashed line, labeled different, is $P_{\text{different}}(t)$, the probability of observing each of the three other states (Equation (3.4)).

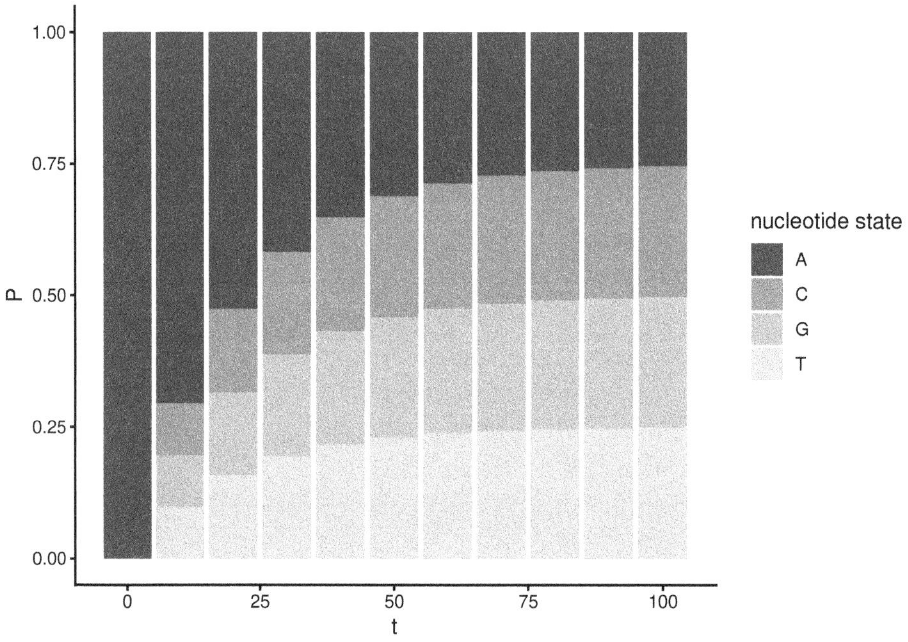

Figure 3.9: Stacked bar plots indicating the frequency of each nucleotide after evolution for a specified amount of time. The rate of evolution is $\mu = 0.050$. The starting state is set at A, so the probability of observing an A is $P_{\text{same}}(t)$ (Equation (3.3)). The other three nucleotides, C, G, and T, are each described by $P_{\text{different}}(t)$ (Equation (3.4)). At time $t = 0$ (no evolution), the probability that the state is the same as at the start is 1.0. As the length of time increases, the four nucleotides converge on equal probability of 0.25 each.

Let's put this back into a biological context. Our simple model allows us to calculate the probability of a given nucleotide state at the end (child node) of a branch, either $P_{\text{same}}(t)$ or $P_{\text{different}}(t)$ depending on whether that state matches the start state, given:

- The nucleotide state at the beginning (parent node) of the branch
- Replacement rate μ
- Length of the branch t

This is powerful stuff. We could do a variety of things with this model machinery, for example:

- Simulate evolution along a branch by sampling nucleotides for the child node from this probability distribution
- Ask the probability of a given starting state given an end state
- Evaluate how reasonable our μ model parameter value is. If, for example, we have a tree with very short branches that had different states at their parents and children, we might be skeptical of a low μ value.

3.3 Generalizing the simple model

The model we built above only has one parameter that can vary, μ, so we can describe the model parameters very simply. This is convenient, but leaves some things happening under the hood a bit opaque. Let's rewrite this simple model to make it clearer how we are using this parameter, and to reveal some other parameters that are there but that we ignored until now because they were clamped. First, we need to represent the model as a 4×4 rate matrix, which we will call $\mathbf{Q}$, as defined in Equation (3.5). Each row corresponds to one of the four possible nucleotides (A, C, G, T, in that order from top to bottom), and each column corresponds to one of the four possible nucleotides (A, C, G, T, in that order from left to right). Each of the elements in the matrix is the instantaneous rate of change from the nucleotide of the corresponding row, to the nucleotide of the corresponding column.

$$\mathbf{Q} = \begin{pmatrix} -3\mu\pi & \mu\pi & \mu\pi & \mu\pi \\ \mu\pi & -3\mu\pi & \mu\pi & \mu\pi \\ \mu\pi & \mu\pi & -3\mu\pi & \mu\pi \\ \mu\pi & \mu\pi & \mu\pi & -3\mu\pi \end{pmatrix} \tag{3.5}$$

Recall that μ is the rate of *any* replacement event happening. That replacement event could be an A, C, G, or T. Only three of these replacements lead to a substitution, since replacing with the original nucleotides does not lead to a change. To find the rate of *specific* replacements happening, as we need to do for the elements of this matrix, we need to apportion the total replacement rate μ to specific nucleotides. We can do that with a new term π, which is the name we will give to the equilibrium frequency of each state. This corresponds to the frequency of each nucleotide in the bag we randomly sampled from. In our simple model, $\pi = 0.25$ for all nucleotides. Because π was clamped and wasn't free to vary, it was essentially invisible in the way we previously described the model.

The off-diagonal elements of $\mathbf{Q}$ give the rates of substitutions, and are all $\mu\pi$. But what's up with the diagonal elements? We pick these diagonal elements to be whatever value leads the rows to sum to 0. The basic intuition of this is that we aren't creating or destroying nucleotides, just replacing them. So the net change needs to be 0. Since there are three possible substitutions in each row, and each substitution has rate $\mu\pi$, these diagonal elements are set to $-3\mu\pi$. The negative rates for the diagonal elements can be thought of as a rate of leaving the current state, while the positive off diagonal rates correspond to entering new states.

There is a lot going on in $\mathbf{Q}$. To make sense of it all, it helps to factor it out into two parts (Swofford et al., 1996). The first is a 4×4 matrix $\mathbf{R}$, which has all the rates, and the second is a 4×4 matrix $\mathbf{\Pi}$ that has the equilibrium frequencies on its diagonal and 0 everywhere else (Equation (3.6)).

$$\mathbf{Q} = \mathbf{R\Pi} = \begin{pmatrix} -3\mu & \mu & \mu & \mu \\ \mu & -3\mu & \mu & \mu \\ \mu & \mu & -3\mu & \mu \\ \mu & \mu & \mu & -3\mu \end{pmatrix} \begin{pmatrix} \pi & 0 & 0 & 0 \\ 0 & \pi & 0 & 0 \\ 0 & 0 & \pi & 0 \\ 0 & 0 & 0 & \pi \end{pmatrix} \tag{3.6}$$

(See Section A.3 for resources on linear algebra if you are unfamiliar with the intuition and mechanics of matrix multiplication.)

$\mathbf{Q}$ is the instantaneous rate matrix—it specifies the particular amount of change we expect over a short period of evolutionary time. But as we discussed before, we often want to know the probability of ending with a particular state if you start with a particular state and let it evolve along a branch of a given length. Before, when we were keeping things as simple as possible, we used the exponential functions $P_{\text{same}}(t)$ and $P_{\text{different}}(t)$ (Equations (3.3) and (3.4)) for this. They took as input the overall replacement rate μ and the length of the branch t. Now we want a similar equation, but we want to provide the rate matrix Q rather than the single parameter μ. Again we can just use an exponential function, and it actually has a much simpler form.

$$\mathbf{P}(t) = e^{\mathbf{Q}t} \tag{3.7}$$

Raising e to the power of a matrix is known as matrix exponentiation, and it returns a matrix with the same dimensions as the matrix in the exponent. This new matrix $\mathbf{P}(t)$, known as the substitution probability matrix, is therefore also a 4×4 matrix. As for $\mathbf{Q}$, $\mathbf{\Pi}$, and $\mathbf{R}$, each row and column corresponds to one of the four possible nucleotides (A, C, G, T). Each of the elements in the matrix is the probability of going from the nucleotide of the corresponding row to the nucleotide of the corresponding column over a branch of length t. Given the single rate parameter and uniform equilibrium frequencies, the diagonal elements are each $P_{\text{same}}(t)$ (Equation (3.3)), and the off diagonal elements are each $P_{\text{different}}(t)$ (Equation (3.4)).

This was a lot of work to write a really simple model in a much more complicated way. Now we can start to reap the rewards of describing this simple model in this form.

3.4 Expanding the models

The mathematical, statistical, and computational machinery above describes the evolution of one DNA site along one branch in a phylogeny (Figure 3.3) according to a simple model. In fact, this is the simplest possible model of DNA evolution, named JC69 after the folks who described it in 1969 (Jukes and Cantor, 1969). It has only one free parameter—the rate of evolution μ. It is highly idealized—there are many important facts about DNA and DNA evolution that we deliberately omit.

The simplicity of JC69 makes it a useful starting place to understand how models of DNA evolution work, but it is too simple to be very useful in practice. There are many other biological aspects of sequence evolution we might want to consider, including that:

- The nucleotides C, G, T, and A are often not found at an equal frequency of 0.25 each. The GTR model lets all four equilibrium frequencies differ, subject only to the constraint that they sum to 100%. It is tempting to think that double-stranded DNA forces C and G to occur at the same frequency, and likewise T and A, so that the four could always be summarized by a single number. That equality is exact only when you count both strands of the duplex together, because every C on one strand is paired with a G on the other (this is Chargaff's first parity rule). But sequence data, and the π parameters of our model, describe a *single* strand. Within one strand the two frequencies are only approximately equal (Chargaff's second parity rule), and real genomes show measurable GC and AT skews, which is why GTR estimates all four frequencies rather than one. As a convenient approximation we can nonetheless summarize a strand by its GC content—the percent of sites that are G or C—by assuming C=G and T=A. For example, the human genome has a GC content of 41%, which under this approximation gives nucleotide frequencies of 0.205 G, 0.205 C, 0.295 T and 0.295 A. This is quite different than the equal frequencies of 0.25 for all nucleotides expected by the JC69 model.
- The rates of mutation between different nucleotides are not the same. This is in part because some nucleotides are shaped more like others, and are more likely to be substituted for each other by mistake. In particular, A and G tend to be substituted for each other because they have two rings, and C and T tend to be substituted for each other because they have one ring. These changes between A/G and C/T are referred to as transitions. All other substitutions are called transversions, and are each less likely.

To accommodate each of these deviations from JC69, we need to add parameters to the model to explain the added complexity. There are all kinds of parameters we could add one by one, and in fact that is how the field proceeded in the decades after JC69 was described. Rather than build the models up in that way, though, it is actually easier to describe a general model that these other models, including JC69, are special cases of.

$$\mathbf{Q} = \mathbf{R\Pi} = \begin{pmatrix} - & \mu a & \mu b & \mu c \\ \mu a & - & \mu d & \mu e \\ \mu b & \mu d & - & \mu f \\ \mu c & \mu e & \mu f & - \end{pmatrix} \begin{pmatrix} \pi_A & 0 & 0 & 0 \\ 0 & \pi_C & 0 & 0 \\ 0 & 0 & \pi_G & 0 \\ 0 & 0 & 0 & \pi_T \end{pmatrix} \tag{3.8}$$

(To keep things compact, I have followed the convention of placing $-$ in the diagonal elements, which is short for "whatever it takes to make the rows of $\mathbf{Q}$ sum to 0".)

This model (Equation (3.8)) is called the General Time Reversible (GTR) model of DNA sequence evolution (Tavaré, 1986). It is General because it has parameters that allow many things to vary independently. The $\pi_A ... \pi_T$ parameters allow the equilibrium frequencies to differ for each nucleotide. The $a ... f$ parameters adjust μ so that rates can differ. For example, the instantaneous rate of change from A to C can be different from that of A to G by setting a and b to different values. It is Time Reversible because we don't have a different parameter for every single off-diagonal element, but instead mirror them across the diagonal: the entry of $\mathbf{R}$ shared by A to C and C to A is μa, the entry shared by A to G and G to A is μb, and so on. It is important to be careful here about what is shared. The mirrored quantity μa is an entry of $\mathbf{R}$, sometimes called an *exchangeability*, not the instantaneous rate of change itself. Recall from Equation (3.5) that the actual rate is the exchangeability multiplied by the equilibrium frequency of the *destination* nucleotide. So the instantaneous rates in the two directions are $q_{A \to C} = \mu a \, \pi_C$ and $q_{C \to A} = \mu a \, \pi_A$, which are equal only in the special case $\pi_A = \pi_C$. What time reversibility actually guarantees is *detailed balance*, $\pi_A \, q_{A \to C} = \pi_C \, q_{C \to A}$: at equilibrium the flow of probability from A to C exactly balances the flow from C to A, even though the per-lineage rates in the two directions differ. There are a few motivations for mirroring the exchangeabilities in this way. It keeps the total number of parameters down. It also reflects what is observed biologically. And it means that we don't need to know the direction of time along a branch to calculate $\mathbf{P}(t)$. This last point is very important since we often don't know where the root of a tree is when we want to calculate these probabilities. It also turns out to be convenient and efficient for phylogenetic inference software tools to reroot phylogenies without changing these probabilities as they make calculations on phylogenies.

If we let all the parameters in the GTR model (Equation (3.8)) be free, it is quite complex. We would have to estimate them all from data. The alternative is to clamp some of them. In fact, by comparison of Equations (3.6) and (3.8), we can see that JC69 is a clamped version of GTR where $\pi_A = \pi_C = \pi_G = \pi_T = 0.25$ and $a = b = c = d = e = f = 1$. Only μ is left free in JC69.

There are a variety of other commonly used models that clamp these parameters in different ways, some leaving more freedom than others. Some that are widely used and have very specific biological motivation have their own names, like JC69 (the motivation for this model is that it is so simple). Examples of other named models include:

- HKY85 (Hasegawa et al., 1985) differs from GTR in setting some rate modifiers equal to each other so that there are two rates, $b = e$ for transitions (A to G, and C to T) and $a = c = d = f$ for transversions (all other changes). It still lets all the π parameters vary independently, unlike the earlier two-parameter model of Kimura (1980), which also distinguishes transition and transversion rates but holds the base frequencies equal.
- F81 (Felsenstein, 1981) differs from HKY85 in that it sets all the rate modifiers to 1, so that $a = b = c = d = e = f = 1$. It is almost as simple as JC69, except that the π parameters vary independently.

We now have a more complex model of DNA evolution that we can simplify by setting the parameters to equal each other and/or clamping them to specific values. It can accommodate much of the variation observed in actual DNA sequence data. We could use it to simulate data just as we did in Figure 3.7, but taking into consideration things like unequal nucleotide frequencies and differences in rates of change between nucleotides. This is getting us much closer to something that provides real biological insight.

3.5 Plugging some numbers into the expanded model

Now that we have a framework for specifying a model with unequal equilibrium frequencies and different rates of change between different nucleotide pairs, let's build a specific case of a model and use it to simulate some data. That means we need to pick some actual model parameters, plug them into $\mathbf{R}$ and $\mathbf{\Pi}$ to get $\mathbf{Q}$, and then exponentiate that to get $\mathbf{P}(t)$. We can then use that to simulate changes in a single nucleotide along a branch. This will be a mathematically grounded version of the schematic in Figure 3.3B.

We are using a mammal tree, so let's pick some parameter values that roughly approximate what we see in mammals. Rather than set all the parameters independently, let's set up an HKY85 model, which accommodates non-uniform base frequencies and different transition/transversion ratios. First, we can clamp $\mu = 1$. Transitions (captured by parameters b and e) are on the order of 4 times more frequent than transversions (captured by parameters a, c, d, and f) in mammals (Rosenberg et al., 2003). So we will clamp $b = e = 2$ and $a = c = d = f = 0.5$. I picked these particular values (rather than others, such as 4 and 1) because they keep the mean of the off-diagonal entries in $\mathbf{R}$ to 1.

Now we have the following values for $\mathbf{R}$:

$$\mathbf{R} = \begin{pmatrix} - & 0.5 & 2 & 0.5 \\ 0.5 & - & 0.5 & 2 \\ 2 & 0.5 & - & 0.5 \\ 0.5 & 2 & 0.5 & - \end{pmatrix} \tag{3.9}$$

We won't specify the diagonal elements of $\mathbf{R}$ quite yet.

Now we need $\mathbf{\Pi}$. We noted earlier that humans have a GC content of 41%, so we'll use that. This gave the following entries for $\mathbf{\Pi}$:

$$\mathbf{\Pi} = \begin{pmatrix} 0.295 & 0 & 0 & 0 \\ 0 & 0.205 & 0 & 0 \\ 0 & 0 & 0.205 & 0 \\ 0 & 0 & 0 & 0.295 \end{pmatrix} \tag{3.10}$$

Now we can calculate $\mathbf{Q}$ by multiplying $\mathbf{R}$ and $\mathbf{\Pi}$ and adjusting the diagonal so that the rows each sum to 0:

$$\mathbf{Q} = \mathbf{R\Pi} = \begin{pmatrix} -0.6600 & 0.1025 & 0.4100 & 0.1475 \\ 0.1475 & -0.8400 & 0.1025 & 0.5900 \\ 0.5900 & 0.1025 & -0.8400 & 0.1475 \\ 0.1475 & 0.4100 & 0.1025 & -0.6600 \end{pmatrix} \tag{3.11}$$

This is our instantaneous rate matrix. From here on I will switch from mathematical notation to computational representations, since we are doing more calculations. Here is what the contents of $\mathbf{Q}$ look like when outputted from R:

```
##         A       C       G       T
## A -0.6600  0.1025  0.4100  0.1475
## C  0.1475 -0.8400  0.1025  0.5900
## G  0.5900  0.1025 -0.8400  0.1475
## T  0.1475  0.4100  0.1025 -0.6600
```

Now we need to exponentiate $\mathbf{Q}$ to get $\mathbf{P}(t)$, according to Equation (3.7). If we do that numerically, we need to specify t before we do the exponentiation. So, for example, $\mathbf{P}(1)$ is:

```
##           A          C          G         T
## A 0.5913935 0.08066121 0.21187182 0.1160735
## C 0.1160735 0.49837662 0.08066121 0.3048887
## G 0.3048887 0.08066121 0.49837662 0.1160735
## T 0.1160735 0.21187182 0.08066121 0.5913935
```

The sum of each row of $\mathbf{P}(t)$ is 1 for any value of t, since our model doesn't create or destroy nucleotides. No matter which of the four nucleotides you start with, you have to end with one of the four nucleotides. This better demonstrates why we had to select diagonal values of the *rate* matrix $\mathbf{Q}$ so that each row in that matrix sums to 0—if we hadn't, then the row sums of $\mathbf{P}(t)$ would not be 1.

Now that we can derive $\mathbf{P}(t)$, let's explore some of its properties. Let's first think about what happens when $t = 0$. In this case, there is no time for evolution to occur and the resulting nucleotide should be the same as the starting nucleotide. Indeed, we see this is the case:

```
##   A C G T
## A 1 0 0 0
## C 0 1 0 0
```

```
## G 0 0 1 0
## T 0 0 0 1
```

The probability of going from A to A is 1 while the probability of going from A to any other nucleotide is 0. The same goes for all of the other nucleotides, giving ones on the diagonal and zeros everywhere else.

What if the branch is very long, say $t = 100$?

```
##       A     C     G     T
## A 0.295 0.205 0.205 0.295
## C 0.295 0.205 0.205 0.295
## G 0.295 0.205 0.205 0.295
## T 0.295 0.205 0.205 0.295
```

In this case, each row has the exact same values. If you look closely, they should be familiar—they are the equilibrium frequencies from $\mathbf{\Pi}$. This matches what happened in Figures 3.7 and 3.8. As branch length increases, the expected frequencies of each nucleotide deviate further and further from the original state and approach the equilibrium frequencies specified by $\mathbf{\Pi}$.

3.6 Simulation along a single branch

We can now calculate the probability of a specific end state given a start state. Now let's use this machinery to simulate evolution along a single branch at a time, as in Figure 3.3B.

We first need to set the state at the parent node. There are a couple of approaches we could use to set the parent state. We could clamp it to a specific initial state, just as we did in our JC69 simulations when we set the starting state to A (Figure 3.4). Alternatively, we could pick the starting state at random.

How do we select a starting state at random? We could draw the starting state from a bag with equal frequencies of each nucleotide (as we did at the beginning of this chapter), but our model allows us to make a more informed selection than that. We implemented $\mathbf{\Pi}$ because we wanted to describe cases where the nucleotides do not occur at uniform frequencies, so let's draw from that distribution instead. For the toy mammal model we made above, that is 0.205 G, 0.205 C, 0.295 T and 0.295 A. We just sample a single nucleotide from this probability distribution.

Table 3.1: Each row is simulated evolution along a single edge.

Parent	Length	Child
A	0.0387691	A
A	0.3444866	A
C	0.6561006	A
C	0.6723020	C
A	0.5311025	A

Parent	Length	Child
A	0.7477842	G
T	1.7042348	T
A	0.6896113	G
G	0.7347406	G
G	1.3263353	G

In Table 3.1 I selected the Parent nucleotide by sampling from $\mathbf{\Pi}$ as described above. I then created a random branch length t by sampling from a uniform distribution that ranges from 0 to 2, just to get a variety of branch lengths. I then simulated the Child state by sampling from $\mathbf{P}(t)$ given t and the Parent state.

3.7 Scaling from a single branch to a whole tree

So far we have considered the evolution of one DNA site along one branch at a time (Figure 3.3B, Table 3.1). We will now expand to a whole tree, keeping our focus on simulation. Our goal is to use the model to simulate the evolution of a single site along all branches, generating a specific nucleotide state at each node. We will use the same toy model of sequence evolution as above. We will consider a simplified tree (Figure 3.10) to keep things compact.

This isn't a big step from what we have already—once we have all the machinery to simulate along a single branch, we can just iterate that to simulate evolution along all the branches in a whole tree.

Let's start with the root of the tree (Figure 3.10, node 5). As in our simulations along single branches, we will pick the state from the equilibrium frequencies $\mathbf{\Pi}$. That gives us the A at the root in Figure 3.10. The root node is the parent of two branches that descend from it. These two branches connect to node 6 (the most recent common ancestor of the clade `(Species_A, Species_B)`) and node 7 (the most recent common ancestor of the clade `(Species_C, Species_D)`). We simulate the states for these child nodes according to the state at the root (node 5), length t of each branch, and $\mathbf{P}(t)$. In each case, this is just as when we simulated evolution along a single branch at a time, it is just that the branches share a parent node so they also share a parent state.

There are four more branches in this tree, each connected to a terminal node. One branch has parent node 6 and child node 1 (which is the `Species_A` terminal node). Now that we are not at the root things are a little different. Rather than draw the state for the parent node from $\mathbf{\Pi}$, we just use the state that was simulated along the branch connecting node 5 to node 6. This state is A. Now we simulate evolution along the branch connecting node 6 to node 1, given the state at node 6, length t of the branch, and $\mathbf{P}(t)$. We then do the same for each of the other branches.

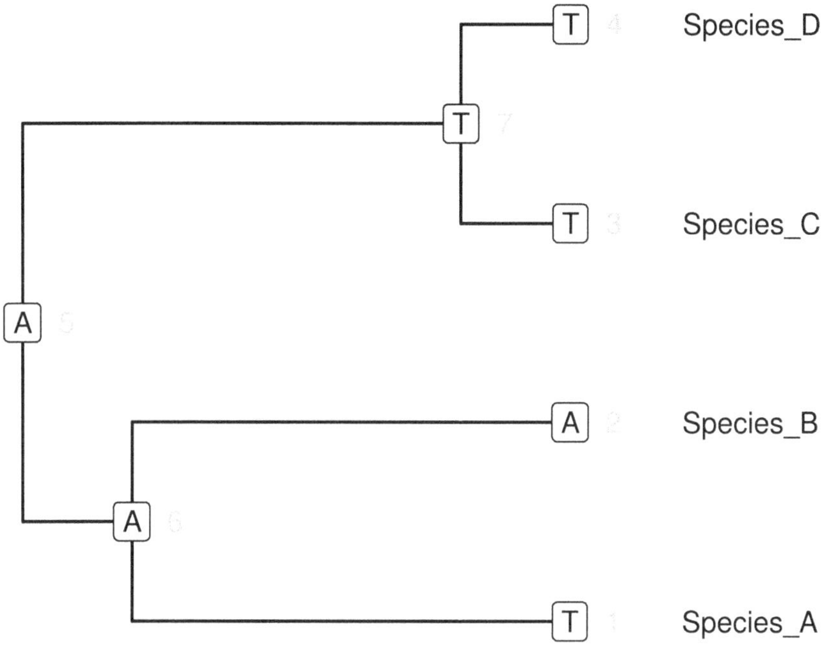

Figure 3.10: Simulation of states for a single DNA site on a simple tree according to our toy mammal model. Node numbers are in gray. Character states are in boxes at nodes. Branch lengths for this phylogram are in units of expected change.

Data can be simulated on a tree of arbitrary size in this way. Just sample the root state from the equilibrium frequencies. Then traverse the tree from the root to each of the tips, simulating the state at each of the other internal nodes and finally the terminal nodes according to the states of their parents, branch length t, and $\mathbf{P}(t)$.

3.8 Scaling from a single site to multiple sites

So far we have considered evolution at a single DNA site across a whole phylogeny. Genomes have from thousands to billions of sites, though. We can trivially expand our simulations from a single nucleotide at a time to arbitrarily long DNA sequences. How? With one simplifying assumption—that the evolution of each site is independent of the evolution at other sites. We just simulate one site at a time, and then stick all the results together into a sequence.

3.9 Concluding thoughts

Here we have built up the conceptual, mathematical, statistical, and computational machinery to simulate DNA evolution on a tree. Sequence simulation is useful for a variety of things, including generating datasets under known conditions to test tools. A major value, though, is to think in a full explicit way about how you are modeling evolution. This probabilistic model framework is the exact same one we will use as we move to our next task, inferring phylogenies from sequence data.

3.10 Additional resources

- As a grad student, I learned much of what I present here from Swofford et al. (1996). This is such a lucid introduction to the likelihood of molecular sequence data on a phylogeny.
- Yang (2014) gives a thorough and rigorous treatment of these substitution models and the statistical framework built on them.
- My own thinking about presenting this material was heavily influenced by Paul Lewis's wonderful lectures at the annual Workshop on Molecular Evolution at Woods Hole. Some of his lectures are now available online as part of the excellent Phylo Seminar at https://www.youtube.com/channel/UCbAzhfySv7nLCrNYqZvBSMg, starting with https://www.youtube.com/watch?v=1r4z0YJq580&t=2111s.
- A great introduction to continuous time models by John Huelsenbeck is available at https://revbayes.github.io/tutorials/dice/.

Chapter 4

Inferring phylogenies from data

In the previous chapter we built modeling machinery that allowed us to specify a rate matrix $\mathbf{Q}$ (Equation (3.8)), and from that derive the matrix $\mathbf{P}(t)$ (Equation (3.7)) that gives the probability of a particular end state given the starting state and the branch length t. We used $\mathbf{P}(t)$ in a generative context, to simulate the evolution of a DNA sequence along the branches of a phylogeny.

We will now turn to using the same statistical framework for what may initially seem to be a very different task—phylogenetic inference, where we infer the topology and branch lengths of a phylogeny from character data (DNA sequences, in this case). But these tasks are similar conceptually. The basic intuition is that you can infer a phylogeny by looking for the topology and branch lengths that are most likely to generate the observed data (Felsenstein, 1981).

This relationship between simulation and inference is widely used in a variety of fields. The probability of the observed data given a hypothesis is referred to as the likelihood. Searching for the most likely hypothesis is referred to as maximum likelihood (ML). A comprehensive reference on the whole endeavor of phylogenetic inference is Felsenstein (2004).

4.1 Probability of a single history

We will consider the toy phylogeny, along with its tip states, shown in Figure 4.1.

We will start by calculating the probability of a single history of evolution for a single site on a single phylogeny with specified topology and branch lengths. This history is a full set of states at all nodes. These are added to the toy phylogeny in Figure 4.2. This isn't a history we have any particular reason to believe; it is just one possible history of states randomly chosen from all the possible histories.

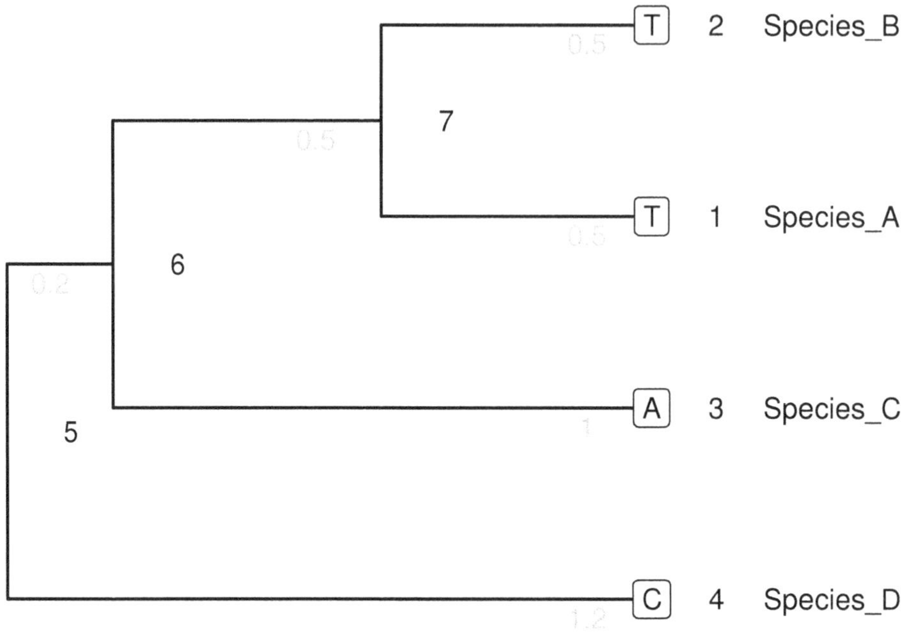

Figure 4.1: The toy phylogeny we will use to examine inference. Node numbers are in black. Branch lengths are gray numbers below branches. Tip node states are within boxes.

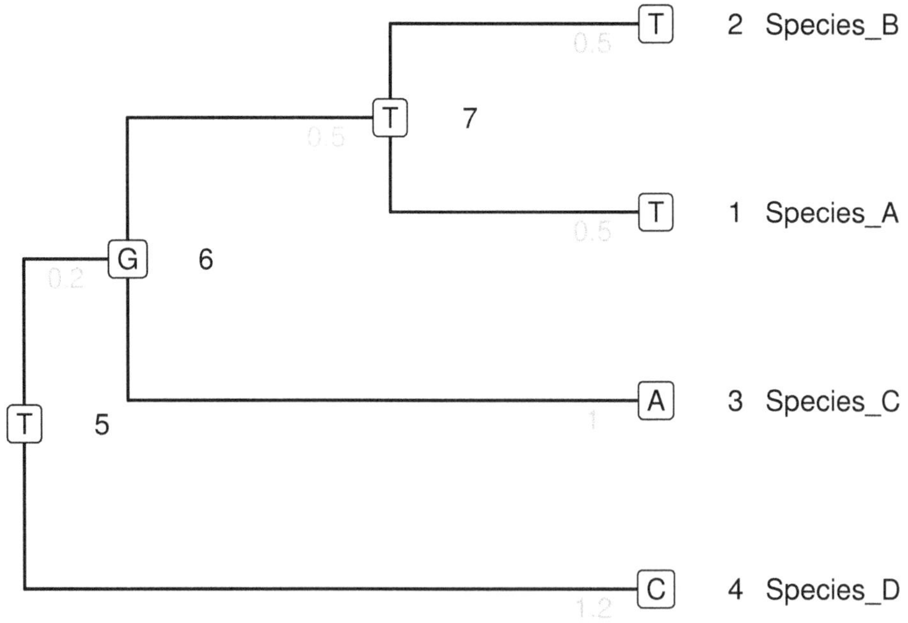

Figure 4.2: The same toy tree as above, now with arbitrary states added at the internal nodes, so every node has a state (in boxes).

Our goal now is to calculate the probability of each observed change. Recall that the matrix that contains these probabilities, given a starting state (rows), ending state (columns), and branch length t, is given by:

$$\mathbf{P}(t) = e^{\mathbf{Q}t} \tag{4.1}$$

Where $\mathbf{Q}$ is the rate matrix. Let's plug some numbers in using the model we specified in the previous chapter. We don't have any specific reason to use this model on this tree; we are just sticking with it since we already built it.

Here is the relative rate matrix $\mathbf{R}$:

```
##     A   C   G   T
## A 0.0 0.5 2.0 0.5
## C 0.5 0.0 0.5 2.0
## G 2.0 0.5 0.0 0.5
## T 0.5 2.0 0.5 0.0
```

The equilibrium frequencies $\mathbf{\Pi}$:

```
##       A     C     G     T
## A 0.295 0.000 0.000 0.000
## C 0.000 0.205 0.000 0.000
## G 0.000 0.000 0.205 0.000
## T 0.000 0.000 0.000 0.295
```

And their product $\mathbf{Q}$, with the diagonal adjusted so that rows sum to 0:

```
##         A       C       G       T
## A -0.6600  0.1025  0.4100  0.1475
## C  0.1475 -0.8400  0.1025  0.5900
## G  0.5900  0.1025 -0.8400  0.1475
## T  0.1475  0.4100  0.1025 -0.6600
```

For each branch, we can now use $\mathbf{P}(t)$ to calculate the probability of a change from the start state at the parent node to the end state at the child node, given the branch length t. The results are shown in Figure 4.3.

Now that we have the probabilities of each of these changes, we can calculate the joint probability of all these changes. When we want to calculate the joint probability of multiple independent events, we take the product of the probability of each specific event. For example, the probability of rolling a 4 on a fair die is $1/6$. The probability of rolling two 4s on two fair dice is $1/6 \times 1/6 = 1/36$. So we can take the product of all the probabilities to calculate the joint probability of all of these events happening. We can think of the probabilities of specific changes along each branch as the probabilities of the state at each child node given the state at the parent node.

node	probability
1	0.7442034
2	0.7442034

node	probability
3	0.3048887
4	0.2260230
5	NA
6	0.0195083
7	0.0652538

Note, though, that the probability for node 5, the root node, is missing. This makes sense since the root is not the child of any branch, and we calculated the probabilities based on changes along branches. We will therefore assess the probability of the root node state according to $\mathbf{\Pi}$, the equilibrium frequencies. This is the same approach we took when simulating data on a tree. When we fill that in, our full set of probabilities is:

node	probability
1	0.7442034
2	0.7442034
3	0.3048887
4	0.2260230
5	0.2950000
6	0.0195083
7	0.0652538

The joint probability of all these states can now be calculated as the product of each state. This comes out to 1.4332602×10^{-5}. There are multiple ways to think about this probability. One is from a frequency perspective. If we were to simulate character states on this tree, we would expect this full set of character states to occur at a frequency of 14.3 times out of a million simulations.

Here we use much of the same machinery as in the previous chapter, but toward a slightly different end. Rather than use the probability distributions to generate nucleotides in a simulation, we instead calculated the probability of a particular set of nucleotides. These may have seemed like very different tasks at first blush, but as you can now see, their mathematical implementation shares many features.

4.2 Probability of multiple histories

Above we considered the joint probability of a specific set of nucleotide states at all nodes, including both tip nodes and internal nodes. Usually, though, we don't know the internal node states. We don't even know what internal nodes exist, which is why we are trying to infer the phylogeny! Instead we have observed states that we got by sequencing organisms at the tips. We want to clamp these tip states and assess their probability on a particular tree (with

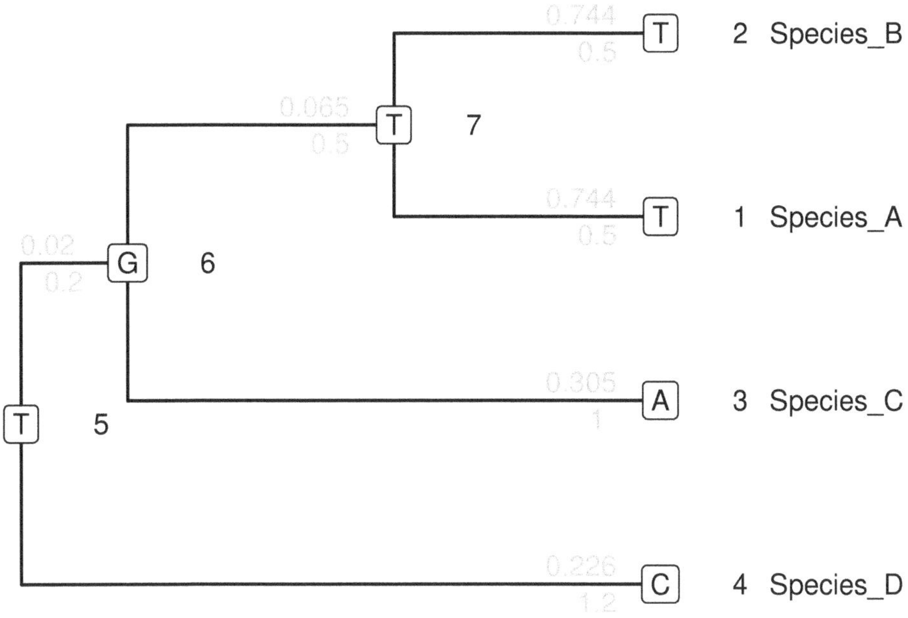

Figure 4.3: The same toy tree as above, but with probabilities of the specific change along each branch above each branch (in gray).

branch lengths) under the model. This probability is independent of a specific history of internal node states.

If we aren't clamping the internal node states as well, how can we calculate the probability of just the tip node states? The key is to consider all possible internal states. Each configuration of internal node states represents one possible history that gave rise to the observed tip states. We can sum the probabilities of each of these different ways to get the tip states to find the probability of the tip states over all possible histories. We are summing the probabilities because these are mutually exclusive histories that could give rise to the observed data. For example, if we want to find the probability of getting a total of seven when rolling two dice, we need to add up the probability of each way to get seven (1+6 *or* 2+5 *or* 3+4 *or* ... 6+1). This is different from when we multiplied probabilities to find the joint probabilities of multiple events occurring together (*e.g.*, the probability of rolling a 4 *and* another 4).

This is a small tree, with only 3 internal nodes that can each have 4 states. This gives $4^3 = 64$ possible histories. That is small enough to list them out below. I also include the probability of each specific history, calculated as I did above (the example above corresponds to row 60 here).

n1	n2	n3	n4	n5	n6	n7	probability
T	T	A	C	A	A	A	0.0000451
T	T	A	C	C	A	A	0.0000048
T	T	A	C	G	A	A	0.0000036

n1	n2	n3	n4	n5	n6	n7	probability
T	T	A	C	T	A	A	0.0000035
T	T	A	C	A	C	A	0.0000000
T	T	A	C	C	C	A	0.0000025
T	T	A	C	G	C	A	0.0000000
T	T	A	C	T	C	A	0.0000002
T	T	A	C	A	G	A	0.0000005
T	T	A	C	C	G	A	0.0000005
T	T	A	C	G	G	A	0.0000044
T	T	A	C	T	G	A	0.0000004
T	T	A	C	A	T	A	0.0000000
T	T	A	C	C	T	A	0.0000003
T	T	A	C	G	T	A	0.0000000
T	T	A	C	T	T	A	0.0000019
T	T	A	C	A	A	C	0.0000282
T	T	A	C	C	A	C	0.0000030
T	T	A	C	G	A	C	0.0000023
T	T	A	C	T	A	C	0.0000022
T	T	A	C	A	C	C	0.0000018
T	T	A	C	C	C	C	0.0002699
T	T	A	C	G	C	C	0.0000013
T	T	A	C	T	C	C	0.0000164
T	T	A	C	A	G	C	0.0000012
T	T	A	C	C	G	C	0.0000011
T	T	A	C	G	G	C	0.0000097
T	T	A	C	T	G	C	0.0000008
T	T	A	C	A	T	C	0.0000006
T	T	A	C	C	T	C	0.0000069
T	T	A	C	G	T	C	0.0000004
T	T	A	C	T	T	C	0.0000432
T	T	A	C	A	A	G	0.0000088
T	T	A	C	C	A	G	0.0000009
T	T	A	C	G	A	G	0.0000007
T	T	A	C	T	A	G	0.0000007
T	T	A	C	A	C	G	0.0000000
T	T	A	C	C	C	G	0.0000018
T	T	A	C	G	C	G	0.0000000
T	T	A	C	T	C	G	0.0000001
T	T	A	C	A	G	G	0.0000017
T	T	A	C	C	G	G	0.0000016
T	T	A	C	G	G	G	0.0000142
T	T	A	C	T	G	G	0.0000011
T	T	A	C	A	T	G	0.0000000
T	T	A	C	C	T	G	0.0000002
T	T	A	C	G	T	G	0.0000000
T	T	A	C	T	T	G	0.0000013
T	T	A	C	A	A	T	0.0005139

n1	n2	n3	n4	n5	n6	n7	probability
T	T	A	C	C	A	T	0.0000552
T	T	A	C	G	A	T	0.0000415
T	T	A	C	T	A	T	0.0000400
T	T	A	C	A	C	T	0.0000071
T	T	A	C	C	C	T	0.0010512
T	T	A	C	G	C	T	0.0000050
T	T	A	C	T	C	T	0.0000638
T	T	A	C	A	G	T	0.0000214
T	T	A	C	C	G	T	0.0000198
T	T	A	C	G	G	T	0.0001776
T	T	A	C	T	G	T	0.0000143
T	T	A	C	A	T	T	0.0000366
T	T	A	C	C	T	T	0.0004512
T	T	A	C	G	T	T	0.0000255
T	T	A	C	T	T	T	0.0028111

Note that I listed the states for all the nodes, including nodes 1-4, which are clamped. It is the last three internal nodes (n5-n7) that have variable states. The probabilities for each specific history range quite widely, from 8.2902428×10^{-9} to 0.0028111.

The sum of the probabilities for each of these different histories for n5-n7 that give rise to the observed clamped states for tip nodes n1-n4 is 0.0058252. This probability of the data given a particular hypothesis (the topology, branch lengths, model, and model parameters) is the likelihood.

4.3 Log likelihood

The likelihood of these data on this phylogeny, 0.0058252, is not a big number. And this is a very small tree. As trees get larger there are many more probabilities we need to multiply, so the products get even smaller. The joint probabilities, in fact, get so small that computers cannot represent them accurately, underflowing toward zero. Rather than store and manipulate the small probabilities directly, most tools take the natural logs of the probabilities, $\ln(p)$. The log likelihood for this phylogeny is -5.1455597. Taking the log transforms probabilities to a numerical representation that is easier to work with. It also has the added value of making calculations of joint probability simpler. Given the relationship between the log of products of variables and the sum of logs of each value:

$$\ln(a) + \ln(b) = \ln(ab) \tag{4.2}$$

We can calculate joint log probabilities as sums of log probabilities for each event, rather than as the log of products of the probabilities. You will almost always see log likelihoods, rather than just likelihoods, published in the literature. Note that because likelihoods are probabilities and therefore range from 0–1, the log likelihoods will range from $-\infty$ (for probabilities very close to 0) to 0 (for probabilities close to 1). Since likelihoods tend to be small, they end up as log likelihoods that are negative numbers with large absolute values.

4.4 Likelihood for multiple sites

The machinery above gives us everything we need to calculate the log likelihood of a specific pattern of nucleotides across tips for a single site in a DNA sequence. We now need to expand this model from a single site to multiple sites within a gene or even across whole genomes.

This comes down to more of the same. We do everything we did above for each site, and then sum the log likelihoods across sites. This gives us the log of the joint probability of observing the data at all sites. This joint probability for all sites will be much smaller than the probability for each individual site.

4.5 Maximum likelihood

At this point we can calculate the log likelihood for specified phylogenies, models, and DNA sequences. But we set out to do phylogenetic inference, where we estimate phylogenies from sequences at tips. How do we get there from here? Once we can calculate the likelihood of a given phylogeny, we can calculate the likelihood of any phylogeny. We can then search for the phylogeny with the maximum likelihood (and, of course, maximum log likelihood).

The small toy phylogeny considered here (Figure 4.1) has four tip nodes. By reference to Equation (2.1), we can see that there are 15 possible topologies. For each, we can optimize the branch lengths to find the maximum likelihood for the topology. This is an iterative process, where each branch length is progressively refined until no change increases the likelihood. An excellent interactive visualization allows you to manually optimize branch lengths on a small phylogeny—http://phylo.bio.ku.edu/mephytis/brlen-opt.html. Then we pick the topology with the maximum likelihood. This requires a very large number of calculations, but is doable for every possible topology.

Things change very quickly, though, as trees grow in size. Beyond about 15 tips there are so many possible topologies that it is impossible to calculate the likelihood for every topology using existing computer hardware and software. That means it is necessary to use heuristics—to modify the tree you have until you can do no better. This is like hill climbing. You calculate the likelihood of a tree and then modify it. If the likelihood is higher, you keep it; if it is worse, you discard it.

This might sound simple, but it isn't. One challenge is that you can get trapped on a local maximum and mistake it for the best phylogeny, when in fact there are other phylogenies with very different topologies that are much better but not locally accessible. There is extraordinary craft that goes into building tools that are able to efficiently climb these likelihood surfaces without getting trapped in local maxima. Widely used maximum likelihood inference programs include RAxML (Stamatakis, 2014) and IQ-TREE (Nguyen et al., 2015).

Optimization of calculations becomes very important. For example, it isn't necessary to recalculate all values on each new topology, since some of the calculations from previous topologies remain relevant (Felsenstein, 1981).

4.6 Optimality criteria

Here we used likelihood as an optimality criterion to search over treespace, the set of all possible phylogenies, to find the phylogeny that maximizes the criterion. There are other optimality criteria that are used in phylogenetic inference. These include parsimony. In parsimony, the minimum number of changes along branches needed to explain the data at the tips is used as the criterion to evaluate each topology (Fitch, 1971). Optimization proceeds by attempting to identify the topology that requires the fewest changes. This requires far less computational power than likelihood, so searches are faster.

Under some conditions parsimony and likelihood will recover similar topologies, but often they do not. This is because they are doing different things that under many conditions lead to different results (Steel and Penny, 2000). For example, we do not always expect the simplest possible explanation for a given pattern to be the best explanation. If a character has a high rate of evolutionary change, then we expect many changes on a tree rather than the fewest possible. This is accommodated in a likelihood framework.

Chapter 5

Molecular inference in practice

In the last chapter we examined the mathematical and statistical underpinnings of phylogenetic inference with maximum likelihood. In this chapter we examine some of the project design considerations and decisions you will need to make in implementing a phylogenetic analysis. We are sticking with DNA sequence data for now, but we will extend these approaches to other types of data later. This is an introduction to the decisions you will face and some of the approaches available, not a comprehensive explanation of how to implement a study.

5.1 Your question

The first step to an analysis is to clearly define your question. There will be many decisions to make along the way, and a clear question will provide an explicit benchmark for working through these decisions—which decision is best for the question at hand? If you have a vague question it will be much harder to decide which steps are best, and many decisions about trade-offs will essentially be arbitrary.

5.2 Taxon sampling

The tips in a phylogeny, be they genes, species, individuals, or some other evolving entity, are referred to as taxa. One of the first things you will want to consider is the breadth and depth of taxon sampling.

One of the most common mistakes in phylogenetic analyses is to include too few taxa. There are many ways that adding taxa helps:

- Adding taxa breaks up branches in the phylogeny. This often improves the performance of maximum likelihood. If you are only interested in taxa 1...20, it still helps to add in taxa 21...50, even if you end up not paying much attention to where they go. They will often improve your ability to infer relationships among the taxa you are most interested in.

- Adding taxa can help improve model evaluation and model parameter estimation.
- Adding taxa reduces the number of assumptions going into a study. You are almost always inferring a phylogeny because you are uncertain at the outset about evolutionary relationships. If, for example, you include only a couple of species from Clade B because you assume it is monophyletic, this poor sampling may be insufficient to reveal that Clade C is in fact nested within Clade B. With insufficient sampling of Clade B, it may instead appear that the two groups are sister clades.
- Adding taxa provides a more rigorous test of phylogenetic hypotheses. Suppose you are conducting a study to investigate the phylogenetic placement of a specific enigmatic species that is so morphologically distinct from other organisms that it is hard to know where it goes. You think it may fall in Clade A, Clade B, or Clade C, so you densely sample those groups. But in fact it falls within Clade D, and your taxon sampling is such that you can't even evaluate this hypothesis because you have no taxa from Clade D in your study.
- Adding more outgroup taxa often improves rooting. Skimping on outgroup sampling is one of the most common mistakes in project design. The best practice in general is to include the most closely related outgroup taxa possible, and some that are progressively more distantly related. The root and the branches that connect to it are the deepest in the tree, and can be the hardest to infer. It is good to densely sample taxa relevant to the root.

There are costs to adding taxa that must be balanced against the benefits above:

- If you are collecting the data, it costs money to add taxa. That money may be best allocated in other ways.
- Adding taxa can considerably increase the computational demand of analyses. This can increase the time and cost of analysis, or even make an analysis unfeasible on the computers available to the investigator.

There are some conditions under which you may want to exclude specific taxa, at least from some analyses:

- The data are contaminated, incorrectly identified, or insufficiently documented.
- Preliminary analyses suggest that the taxon has properties very different from other taxa in the study, such as a very different equilibrium frequency or an exceptionally high rate of evolution.
- The taxon has a very high fraction of missing data. Sometimes missing data make a taxon unstable, which can obscure well-supported relationships among other taxa. Sometimes missing data can interact with other factors to lead to bias in the placement of the taxon.

5.3 Collecting sequence data

You will also need to consider which genome regions to include in your analyses. To build a phylogeny, you need homologous characters across the taxa under investigation. In the context of DNA analyses, this means that you need overlapping genome regions across the taxa that will be in your phylogeny. Decisions about which specific taxa to include interact with decisions about which genome regions you will consider, so there is often back and forth as you refine these decisions. If you are using publicly available data, your taxon and genome region sampling will be strongly constrained by what data are already available.

You will need to decide if and how to enrich for particular genome regions prior to phylogenetic analyses. Enrichment occurs in two stages. The first is sample enrichment. It takes place at the lab bench after you extract your DNA and before you sequence, and uses molecular methods to bias which DNA regions are sequenced. The second is data enrichment. It takes place after you sequence and before you perform phylogenetic analyses, and uses computational methods to obtain a useful subset of data best suited to your question.

Here I consider a few sample enrichment strategies. I start with methods that use the least sample enrichment since, as sequencing technologies improve and fall in price, these will become more favorable for a given project.

5.3.1 Whole genomes

It is now routine for phylogenetic studies to use shotgun sequencing to sequence entire genomes of sampled taxa without any sample enrichment at all. This is often cheaper and more productive than taking steps to only sequence regions of interest. Studies differ greatly, though, in how these whole genome data are used. In clades with small genomes with minimal sequence divergence, such as some viruses and bacteria, entire genomes can be sequenced and assembled.

For larger genomes, such as plants and animals, data are processed to isolate a subset of homologous regions across taxa. The processing depends on the questions at hand and genomic resources already available. Several strategies include:

- Mapping reads to reference genomes and calling variable sites. If genome assemblies are available for the clade of interest, then reads from each new sample can be mapped to the genome. Sites that are variable across samples can then be used for phylogenetic inference. This makes excellent use of the full dataset, often providing hundreds of thousands of phylogenetically informative sites. It is limited, though, to studies where reference genomes are available and are comparable across the taxa included in the analysis.
- Assembling and analyzing organelle genomes. Sometimes mitochondrial and plastid genomes are sufficient for answering the questions that motivate a study. Because these genomes are small compared to nuclear genomes, they can often be readily assembled from raw sequence reads

and then aligned and analyzed. This approach can be applied regardless of existing genome resources.

- Using other methods to isolate homologous regions across studies. The number of tools that can process raw reads to isolate homologous genome regions across samples without the need for a reference genome is growing. These tools include sharkmer, written by Sam Church and me, and available at https://github.com/caseywdunn/sharkmer, which can produce gene sequences given Polymerase Chain Reaction (PCR) primer sequences and raw DNA sequence data.

Multiple strategies above are often used in the same study; they are not exclusive. Apart from reduced expenses in the lab, one of the values of whole genome sequencing is that it is compatible with so many downstream analysis methods.

5.3.2 Transcriptomes

A transcriptome is the collection of RNA in a cell or collection of cells. Because RNA is transcribed from the genome, sequencing the RNA gives a snapshot of an enriched subset of the genome. Whereas the other enrichment approaches described are implemented by the investigator at the bench, transcriptome sequencing largely relies on a natural process in the living cell (transcription) to enrich for particular genome regions.

In a typical phylogenetic study based on transcriptome data, messenger RNA (mRNA), which is transcribed from protein-coding genes, is isolated. mRNA is then copied to complementary DNA (cDNA) and sequenced. Because many of the same genes are highly expressed across species and tissues, there is considerable overlap between the genes in transcriptome datasets from different samples. Many investigators are interested in protein-coding genes for other purposes as well, adding value to the transcriptome data.

There are a couple of challenges to working with transcriptomes. For one, RNA is much less stable than DNA. It is therefore critical to take considerable care when collecting, storing, and processing tissue for transcriptome studies. Transcriptome sequencing only captures genes that are being expressed, leading to missing data for some genes in some samples. As sequencing costs have fallen, genome sequencing has become a better option than transcriptome sequencing for most phylogenetic projects.

5.3.3 Targeted enrichment

In targeted enrichment, an investigator designs short bait sequences that are similar to conserved genome regions. They then isolate DNA from their taxa of interest, fragment it, and use the baits to fish out the conserved regions along with their neighboring sequences. This allows the investigator to create pools of DNA that are enriched for specific pre-selected regions. This is a highly cost-effective way to isolate homologous sequences across species, and works well when samples have fragmented DNA (in museum specimens, for example).

The baits are typically designed based on one or more genomes that are already available in the clade of interest. There are a few challenges to using targeted enrichment. One is that data are difficult to combine across different studies that used different baits. In addition, the design and synthesis of baits require a large up-front investment that makes this approach best suited to larger projects.

5.3.4 RAD-seq

Like targeted enrichment, RAD-seq enriches for sequence data from specific regions scattered across the genome. It differs, though, in that it uses intrinsic properties of the genome for enrichment, rather than user-designed baits. Genomic DNA is digested with one or two restriction enzymes that cut the genome at specific small sequences, and leave the DNA with particular overhangs at the cut end. These cut fragments are then size-selected. If cutting with two restriction enzymes, the fragments that have different overhangs (because they were cut once with each enzyme) are then further enriched. The fragments are then sequenced.

The result is that sequences are enriched for regions with similar distances between restriction enzyme sites across taxa. This is a powerful way to get a consistent subset of the genome across many different samples. It doesn't require a big investment in baits at the outset, but there are a few important limitations. First, it only works well on closely related taxa, since differences in restriction site placement evolve relatively quickly and enrichment overlap falls quickly with evolutionary distance. RAD-seq therefore is most often used when looking at evolutionary relationships within species or among very closely related species. Second, there is very little control over which specific genome regions are sequenced. For example, if you want to build a phylogeny and look at the evolution of a particular gene, you can't readily engineer the sample enrichment to target that specific gene.

5.3.5 PCR

The most extreme form of genome enrichment is PCR. This method enriches for a small number of specific genome regions selected by the user. Short DNA oligonucleotide sequences are designed to flank the region of interest and target it for amplification. By placing these oligonucleotides in conserved regions and engineering them to bind slightly different sequences, they can be used to successfully enrich the same genome region from quite distantly related species. PCR enabled the first wave of DNA sequence-based phylogenetic inference starting in the 1980s, and is still widely used today even as a variety of new enrichment methods have become available.

PCR is cheap and relatively easy. Because it has been used for decades to sequence a small number of genes across a wide diversity of organisms, it is a great way to collect data to analyze in combination with publicly available data. Ironically, though, it is now very expensive per nucleotide to collect data with PCR relative to the other methods above. That is because it is difficult to leverage the advantages of modern high-throughput sequencing, which is very cheap per nucleotide, with such a small number of short amplified sequences. As one of the most extreme forms of genome enrichment available, it provides access to a very small fraction of the genome. PCR is still a convenient way to get a tiny amount of sequence data from a large number of taxa.

5.4 Data processing upstream of phylogenetic analyses

Each enrichment strategy requires its own series of data processing steps to get the data ready for phylogenetic inference. These steps vary greatly across enrichment approaches, requiring distinct methods and software tools. The data processing that occurs upstream of the actual phylogenetic analysis has two primary objectives: estimating the sequence of the sampled genome regions, and enriching for homologous sequences across taxa.

5.4.1 Estimation of original sequences

Raw sequence data have a few properties that make them unusable for direct analysis:

- They are redundant. The same genome regions are sequenced multiple times. PCR products are typically sequenced a couple of times (once in each direction), and high-throughput sequencing often sequences each region hundreds or thousands of times.
- They have many errors. These are introduced during sample preparation at the bench, by the sequencing instruments during data acquisition, and in the initial base calling that generates the sequence file. Errors in raw data vary across sequencing technologies, and are on the order of a few per hundred or thousand nucleotides.
- They are fragmented. Often the raw sequences do not span the full genome region of interest, and must be tiled together to predict longer regions.
- They are ambiguous. It often isn't clear what region of the genome a sequence belongs to, due to genomic repeats, heterozygosity, and genome duplication. This ambiguity must be resolved or eliminated, depending on the application.

The above technical challenges are usually addressed simultaneously in an analysis process called assembly. The goal of assembly is to predict the sequences of the original genome regions based on the raw sequence data. The general idea is to line up overlapping raw sequence data, correct errors by comparing sequences, and collapse all the sequences down to hypotheses about the original actual sequence from which all the raw data were generated. This is a very complex process, and still a major challenge for many genomes and transcriptomes in particular. The best approach differs quite a bit depending on the enrichment strategy, sequencing technology, and the properties of the genome (such as how repetitive it is).

The process of assembling a genome (or genome regions) from scratch is referred to as *de novo* assembly. If high-quality closely related genomes are available, they can be used for guided assembly or mapping, where the raw data are tiled onto the reference rather than assembled from scratch.

5.4.2 Identification of homologous genome regions

It isn't enough to collect and assemble sequence data from the taxa of interest. We also need to collect and identify homologous sequences that represent regions that are shared because they derive from the same region in common ancestors of the taxa.

Like assembly, the optimal approach for identifying homologous sequences depends greatly on the enrichment strategy used. This step is quite straightforward when targeted capture and PCR are used as enrichment strategies, because the data are from specific known regions that were often selected in part because their homology is easier to assess.

In some cases, such as with closely related viruses, genomes can be directly aligned with each other along their full lengths. In larger genomes and transcriptomes, protein-coding sequences are often extracted from the assemblies and are the focus of phylogenetic analyses. There are various approaches to identifying homologous protein-coding genes across taxa. One is to identify sequences for a preselected panel of genes that are selected because they tend to be present in most datasets, they have properties of molecular evolution that are well-suited to phylogenetic analysis, and their homology is thought to be easier to assess. Another is to compare all gene sequences across all species to identify homologous regions, using tools such as OrthoFinder (Emms and Kelly, 2019). This is more computational work, but has the advantage of optimizing gene selection to the data and question at hand.

5.5 Working with publicly available data

Most phylogenetic studies include at least some publicly available data from previously published studies. Some rely entirely on publicly available data, while others add new sequence data as well. The primary repositories for DNA sequence data in published studies, including phylogenetic studies, are NCBI, EMBL-EBI, DDBJ, and NGDC. Many of the same data are available across these platforms.

Most phylogenetic studies deposit their raw data in these public archives, but the raw data can require considerable processing (including assembly) for use in phylogenetic inference. It is often hard to track down assemblies and other intermediate products that are easier to work with since they are not consistently deposited in centralized data archives. Wrangling publicly available data and getting it ready for analysis can be one of the most time-consuming components of a phylogenetic inference project. This is one of the major challenges to conducting phylogenetic analyses of DNA sequence data at this time.

When I am interested in building a phylogeny for a particular group of organisms, one of my first steps is to take a look at the NCBI taxonomy browser. There you can enter a taxonomic group name, and select which data you would like to see (Figure 5.1). You can adjust the number of taxonomic levels to control whether you want to view just summaries for larger groups or get all the way down to the species level.

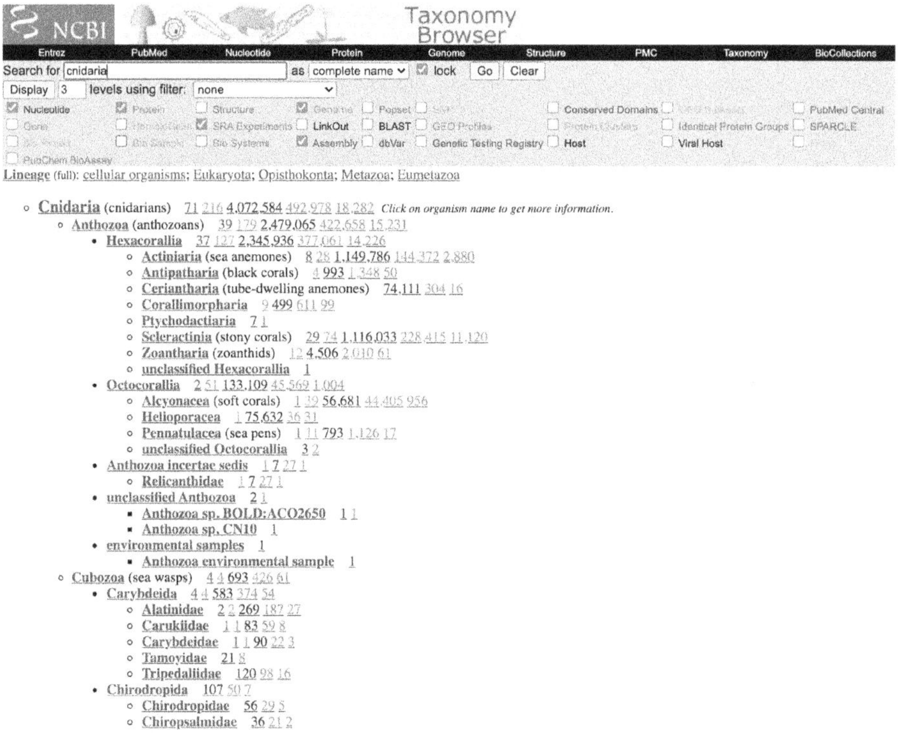

Figure 5.1: A view of the sequence data available for Cnidaria. Here I have selected to view available Nucleotide, Protein, SRA, Genome, and Assembly data.

Based on this overview, you can assess which types of data (*i.e.*, which data produced by which enrichment strategy) are best for advancing your question with publicly available data. All the PCR data and most of the raw sequence data will usually be available here, but many assemblies and processed data for whole genomes, transcriptomes, and other high-throughput datasets may not be. The best practice when working with these data is to take a deep dive into the recent literature and see if these intermediate products were deposited in another public archive, or contact authors for more details.

5.6 Homologous sites

Once you have collected homologous sequences for one or more genome regions across your taxa of interest, you next need to identify homologous nucleotide sites within these regions. This process is referred to as multiple sequence alignment (MSA). Rather than rely on explicit models of molecular evolution, multiple sequence alignment usually relies on similarities between sequences. The general idea is to slide the sequences along each other and insert gaps as needed so that similarity at each site is optimized while reducing the number of gaps introduced. Widely used multiple sequence alignment programs include MAFFT (Katoh and Standley, 2013) and MUSCLE (Edgar, 2004).

The output of multiple sequence alignment is a matrix. Each row is a taxon that will correspond to a tip in the phylogeny. Each column is a single nucleotide site that is hypothesized to be homologous across taxa. There are multiple file formats that are used to store aligned sequences, including FASTA, NEXUS (Maddison et al., 1997), and PHYLIP.

5.7 Combining information across genes

Sometimes when we build phylogenies from genomic data, we are interested in the specific evolutionary history of the genome regions themselves. This is the case, for example, when looking at gene family evolution. Often, though, we are using the evolutionary history of the genome regions as a proxy for the evolutionary history of the organisms, such as when we are asking questions with DNA data about how species are related to each other. There are a variety of conditions, though, under which different genome regions can have genuinely different evolutionary histories from each other, and some of these histories deviate from the evolutionary history of the taxa as a whole (Maddison, 1997). These differences are not errors; they are the expected result of real biological processes, including incomplete lineage sorting, gene duplication and loss, and horizontal transfer. These processes, and the methods that address them, are the subject of Chapter 7.

In cases where such differences are expected to be rare or have little impact, it is common practice to concatenate alignments. Each homologous gene region is aligned independently and then the alignments are joined end to end into a giant superalignment. In this case, information is integrated across regions prior to phylogenetic inference. This is one of several strategies for combining information across regions, which we compare in Chapter 7.

Chapter 6

Evaluation

6.1 Model evaluation

We have examined models of DNA evolution in simulation and inference, but not yet how to select a model. Given a dataset, consisting of aligned DNA sequences for a set of taxa that correspond to tips on a phylogeny, how do you decide which model to use?

6.1.1 Model structure

Each model parameter can be treated in one of three ways (Höhna et al., 2014):

- Constant. The parameter is clamped to a specific value before the analysis and is not free to vary.
- Stochastic. The parameter is free to vary in the analysis. This allows the value to be estimated from the data as part of the inference process.
- Deterministic. The parameter value depends on the values of other parameters according to specified mathematical relationships. Its value can vary, but is determined by the values of other parameters and cannot be set independently from them.

The most widely used DNA sequence evolution models include the General Time Reversible model and its derivatives (Section 3.4). The GTR model has 11 parameters (Figure 6.1). These include the global rate μ used to tune the overall rate of evolution. Next come the relative rate parameters a, b, c, d, e, f that modify the rates of change between particular nucleotide states, so that they can differ from each other. Finally we have the equilibrium frequencies $\pi_A, \pi_C, \pi_G, \pi_T$.

These parameters are treated as follows in the GTR model (Figure 6.1):

- μ is constant. With $\mu = 1$, the branch lengths are in units of expected evolutionary change.

- The six rate parameters are constrained such that $a+b+c+d+e+f=6$. If they were all free to vary, then values that have a sum other than 6 would lead to changes in the global rate rather than the relative rates. Imagine setting them all to 10, for example. This would be equivalent to setting them all to 1 and setting $\mu = 10$. This mathematical relationship between the relative rate parameters means that only five of them can vary independently, because the sixth will depend on the other five and the fact that they all sum to 6. This means that five of the rate parameters are stochastic, and one is deterministic. For our purposes it doesn't matter which one is deterministic, only that one of them is. I'll treat f as the deterministic rate parameter.
- The four equilibrium frequencies are constrained such that $\pi_A + \pi_C + \pi_G + \pi_T = 1$. This is because they are exclusive frequencies, and there are no other possible states. Given that every site must be an A, C, G, or T, their frequencies must sum to 1. This mathematical relationship between the equilibrium frequency parameters means that only three of them can vary independently—three are stochastic and one is deterministic. Again, it doesn't matter which one is deterministic, only how many are stochastic. I'll treat π_A as the deterministic parameter.

The number of stochastic parameters in a model is referred to as the degrees of freedom, df. You can think of it as the number of knobs that can be turned freely during the analysis. Models that have higher degrees of freedom are often referred to as more complex than models with fewer degrees of freedom.

The GTR model has $df = 8$. There are 5 stochastic relative rate parameters and 3 stochastic equilibrium frequencies (Figure 6.1). The other models we have seen are nested within this. By nested I mean that they can take on a smaller subset of the parameter values than the more complex model can. Models that are nested within other models have a smaller degree of freedom. See the IQ-TREE DNA model documentation for a longer list of models—http://www.iqtree.org/doc/Substitution-Models#dna-models.

The HKY85 model has $df = 4$ (Figure 6.1). There is 1 stochastic relative rate parameter, which determines the transition to transversion ratio. There are the same 3 stochastic equilibrium frequency parameters as in the GTR model. There are many values that a GTR model can take that an HKY85 model cannot, for example b can differ from e in GTR but not in HKY85. Every value that HKY85 can take can also be taken by GTR. For example, in HKY85 $b = e$, and in GTR b and e are independent stochastic variables that can be different or that can take on the same value. Because every possible value of HKY85 is also possible in GTR, and GTR has more degrees of freedom than HKY85, HKY85 is nested within GTR.

Likewise, F81 is nested within HKY85 (and therefore GTR as well). It has $df = 3$, the same 3 stochastic equilibrium frequency parameters as in the GTR and HKY85 models. All the relative rates are constant and set to 1. Any value that F81 can take can also be taken by HKY85 or GTR.

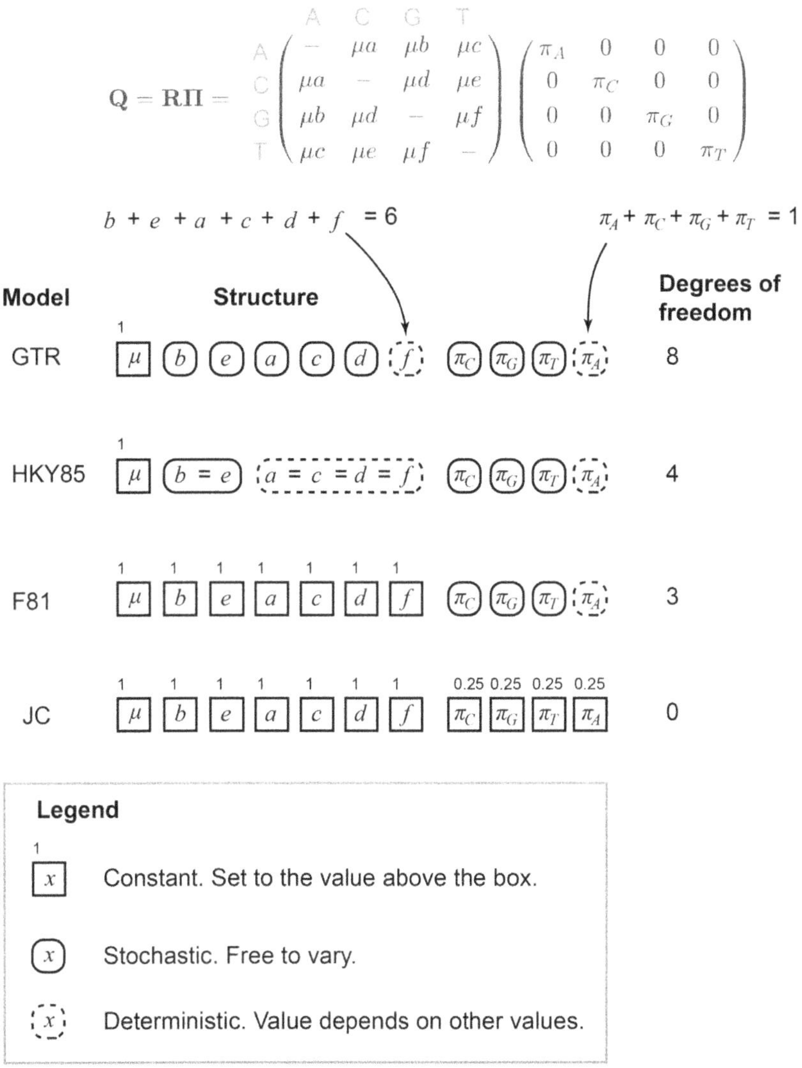

Figure 6.1: A hierarchical view of DNA substitution models. The number of degrees of freedom is determined by the number of independent stochastic parameters (boxes with rounded corners). All other parameters are either constant (set to a specific value ahead of the analysis; boxes with straight lines) or deterministic (their value depends on the value of other parameters according to specified relationships; boxes with dashed lines). Here $\mu = 1$, such that the branch lengths in the phylogeny are the expected amount of evolutionary change. The models are listed from top to bottom by increasing nestedness. Rates are ordered so that transitions and transversions are adjacent. Any model could be realized as a subset of the possible parameter space of the models above it. The visual nomenclature is inspired by Hohna et al. (2014).

The first model we introduced, JC, has $df = 0$. All the relative rates are constant at 1, and all the equilibrium frequencies are constant at 0.25. In typical use, $\mu = 1$. All the other models mentioned above have stochastic parameters that can take on these values, so JC is nested within them all.

Though nested models always have different degrees of freedom, there is no guarantee that models with different degrees of freedom have a nested relationship. It could be that the simpler model has some stochastic parameters that are not stochastic in the more complex model, even though the more complex model has a greater number of free parameters overall. That would allow the simpler model to take on values that the more complex model cannot, violating the nested relationship. When evaluating nestedness it is therefore critical to consider model structure, not just the degrees of freedom.

6.1.2 Evaluating models

In our previous examinations of inference, we used likelihood as an optimality criterion when looking for the phylogeny that best explains our data. Similarly, we can search for the model that best explains the data. We can pick a reasonable phylogeny and evaluate the likelihood of the data under each of the models we would like to consider. In each case we also optimize the model parameter values for each model to make sure we are comparing the models in the best possible light.

We can then make a series of pairwise comparisons between models, where we consider the ratio of their likelihoods. This is the gist of a likelihood ratio test (LRT). Instead of considering the ratio of likelihoods, we can consider the difference in their log likelihoods since:

$$\Delta = \ln\left(\frac{a}{b}\right) = \ln(a) - \ln(b) \tag{6.1}$$

If this difference Δ is positive, then the model corresponding to a is more likely. If it is negative, then the model corresponding to b is more likely.

Let's say we are comparing GTR to HKY85, and we denote the likelihood under GTR as L_{GTR} and the likelihood under HKY85 as L_{HKY85}. We'll put the model with more parameters, GTR in this case, in the numerator, *i.e.*, set a above to L_{GTR} and b to L_{HKY85}.

$$\Delta = \ln\left(\frac{L_{GTR}}{L_{HKY85}}\right) = \ln(L_{GTR}) - \ln(L_{HKY85}) \tag{6.2}$$

One way to proceed would then be to select GTR if $\Delta > 0$, and HKY85 if $\Delta < 0$. This wouldn't be a good way to go, though, since it would pick GTR every single time. The reason is that HKY85 is nested within GTR. That means that the best possible likelihood under HKY85 is also available under GTR. Because it has more degrees of freedom, GTR has many other possible values, and chances are very good that one of those will be more likely than the parameter values that were most likely under HKY85. If comparing two nested models, the model with more degrees of freedom will never have a likelihood lower than the simpler model.

This would seem to imply that more complex models are always better, but they are not. As we add degrees of freedom, we are adding stochastic parameters that must be estimated from the data. In a very extreme case, we clearly couldn't estimate an infinite number of parameters from a finite dataset—there wouldn't be enough information to independently assess each parameter. But there are challenges even with a relatively small number of parameters. If we add too many model parameters, we can overfit. Essentially, if there are too many model parameters we can make any phylogeny look good by adjusting the model parameters. This makes it more difficult to optimize the topology and branch lengths based on their impact on the likelihood. The data have finite information, and the more information we use to estimate model parameters, the less we have to estimate the phylogeny.

When comparing nested models, the question therefore isn't whether one model has higher likelihood than the other, but whether the increase in likelihood one gets from adding parameters is worth the cost of adding the parameters. There are a few different ways to make this cost/benefit analysis.

The simplest is to slightly modify the way we compare the log likelihoods, so that their difference is distributed according to a χ^2 distribution with degrees of freedom equal to the difference in degrees of freedom of the models:

$$\delta = 2\left(\ln(L_1) - \ln(L_0)\right) \tag{6.3}$$

We can compare this test statistic δ to the appropriate χ^2 distribution to assess if the more complex model has a significantly greater likelihood. If not, we stick with the simpler model. It is important to note that it is only appropriate for comparing *nested* models.

There are a few challenges to applying the LRT to model selection in phylogenetics (Posada and Crandall, 2001; Posada and Buckley, 2004). One issue is that the test presumes that one of the models is correct (Zhang, 1999), but in phylogenetics models are always simplifications of the evolutionary process at hand. The impact of violating this assumption varies across datasets and analyses, but can lead to undesirable behavior in real-world applications.

There are two other approaches commonly used in phylogenetic model selection. Where L_i is the likelihood under model i, k_i is the degrees of freedom for the model, and n is sample size (*e.g.*, number of sites in the alignment), these are the Akaike Information Criterion (AIC) (Akaike, 1974):

$$AIC_i = 2k_i - 2\ln(L_i) \tag{6.4}$$

And the Bayesian Information Criterion (BIC) (Schwarz, 1978):

$$BIC_i = k_i \ln(n) - 2\ln(L_i) \qquad (6.5)$$

To apply either of these criteria, they are calculated for each model and the model with the *lowest* value is selected. Adding parameters increases the penalty term, offsetting any increase in likelihood they provide. These criteria have multiple advantages over LRT, including that they can be used to compare non-nested models (Burnham and Anderson, 2002).

To decide whether to use LRT, AIC, or BIC model selection criteria, you need to run a model selection criterion selection analysis. Just kidding. To decide which to use, you should apply your knowledge of phylogenetic methods critically. In many cases, these different approaches will lead to very similar decisions about model selection. As an example of a fairly typical approach, the program IQ-TREE runs AIC and BIC analyses through ModelFinder (Kalyaanamoorthy et al., 2017), but proceeds under the model selected by BIC unless you explicitly step in to apply a different model. If, on inspection of the analysis results, you find that AIC selects a very different model, it would be prudent to run your analyses under that model as well to see if it leads to differences that are relevant to the questions that motivate your project.

6.1.3 Adding complexity to DNA evolution models

The GTR model, and its nested derivatives, capture only a small subset of evolutionary processes and therefore can't describe many of the patterns in observed data. One of the most obvious patterns you will notice when inspecting multiple sequence alignments is that the amount of variation is very different across sites. Some columns will be highly variable, while others are nearly constant. This pattern is due to extensive heterogeneity across sites in rate of evolution.

Two additional model features are often added to address this rate heterogeneity—`I` and `G`. These add additional degrees of freedom. The `I` parameter is the fraction of sites that are invariant and effectively have a rate of zero. `G` accommodates rate heterogeneity across sites that do vary (Yang, 1994). This is modeled with a continuous Γ distribution that is divided into discrete rate categories, usually 4, to speed up computations. Sites are assigned to these different rate categories. To specify the shape of the Γ distribution, and therefore the rates of the categories, a parameter referred to as α is used.

`I` and `G` can be added to GTR and its derivative models. Model selection routines usually examine a panel of models without these rate heterogeneity features, then with `I`, `G`, and `I+G`. The models with `I+G` are often selected, reflecting the importance of accommodating rate heterogeneity in real data.

6.2 Topological evaluation

Maximum likelihood inference gives a point estimate of the phylogeny—a single phylogeny (topology and branch lengths) that maximizes the probability of the observed data under the model. A maximum likelihood analysis will always return a single phylogeny, regardless of how strong the support for that phylogeny is or how much higher its likelihood is than that of other phylogenies. Presenting a phylogeny without any indication of topological support is akin to presenting an estimate of mass made from multiple observations without showing error bars. It is difficult to interpret the point estimate without having a sense of how much uncertainty there is.

6.2.1 Summarizing topological support

Before we get into how to assess support for a phylogenetic hypothesis, it is helpful to think about what that support means and how it is viewed on a phylogeny. Given a specific topology, such as a maximum likelihood phylogeny, how can we summarize and present support for that topology? I'll refer to the topology we are evaluating as the *focal topology*.

Most methods that evaluate topologies generate a large set of phylogenies, which I'll call the *sample*. Assessing support for a feature of the focal topology is a matter of assessing the frequency of phylogenies in the sample that also have that feature. We could assess the support for the focal tree as a whole by asking how frequent identical topologies are in the sample. But support can often be strong in one part of a phylogeny and weak in another. Reporting equivalence of the entire topology provides no window into that variation. This approach would also break down as taxa are added, which quickly increases the number of possible topologies and reduces the chances of any two analyses returning the exact same topologies, given the variation that is inherent in heuristic searches.

Branch frequencies are far more useful. It is helpful to think of a branch as a split (also sometimes called a bipartition) that separates all the taxa in a phylogeny into two groups—those on one side of the branch, and those on the other (Figure 6.2). We will consider two branches to be equivalent if they lead to the same taxon split. This allows us to discuss the equivalence of branches deep in a phylogeny, even when there are many other topological variations elsewhere in the phylogeny.

We can consider each branch in the focal phylogeny independently. For a given focal branch, we count the fraction of phylogenies in the sample that have the same branch, as determined by producing an identical split in taxa. We consider this frequency as the support for the focal branch. If the frequency is 1, the branch was in all the phylogenies in the sample. If it is zero, it wasn't present in any of the phylogenies in the sample. The interpretation of these frequencies, which are often reported as percentages, depends on the method that was used to generate the sample.

It should be noted that these support values are often referred to as "nodal support values", and are often drawn onto nodes. This is unfortunate, as they are branch support values. They are just commonly associated with the child node of the branch rather than the branch itself. This obscures their meaning, and leads to serious problems when trees are re-rooted (Czech et al., 2017).

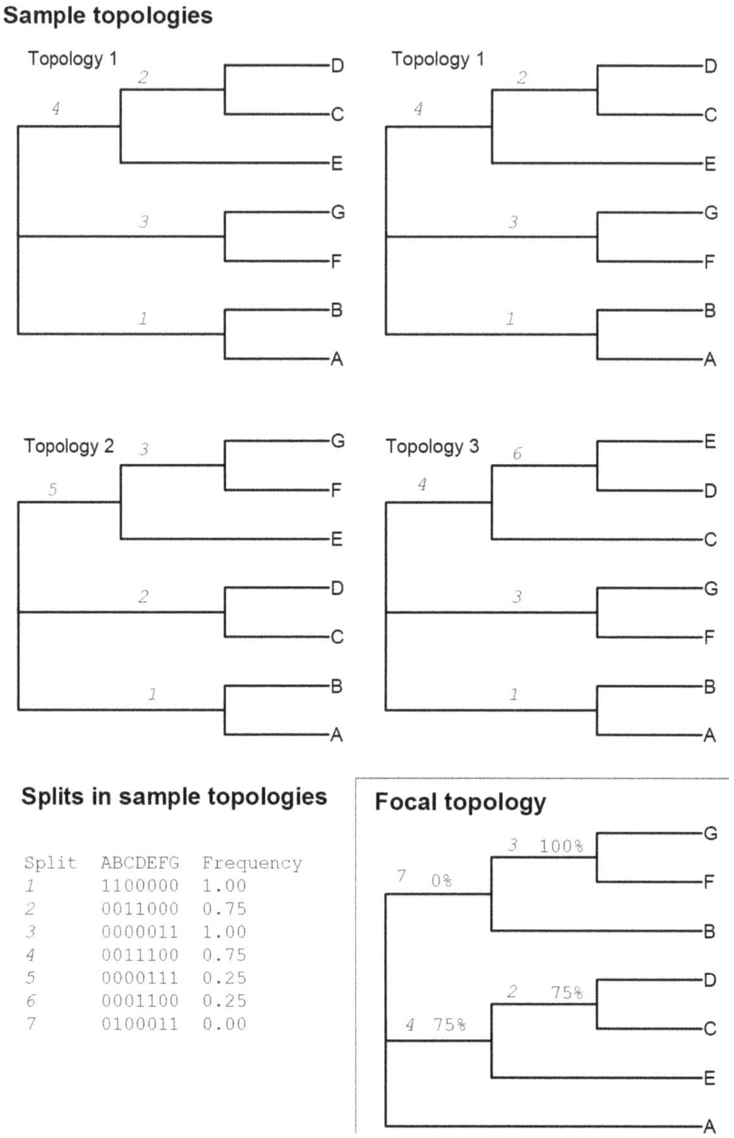

Figure 6.2: Four phylogenies in a sample, one focal phylogeny, and a table of splits found in all of these topologies. A split is a branch, identified by which taxa are split from each other by the branch. The splits table shows the binary encoding of each split, where taxa on the same side of the split have the same binary number (0 or 1). The assignment of 1 or 0 to a particular side of the split is arbitrary. Identical splits are labeled consistently above the branches throughout the figure. The frequency of the split is based on the proportion of sample phylogenies that contain the split. The frequencies of the sample splits are shown as percentages on the branches in the focal topology.

6.2.2 Bootstrapping

The most widely used approach to assessing confidence in maximum likelihood phylogenetic inference is the bootstrap (Felsenstein, 1985a). Given a data matrix where rows are taxa and columns are characters (nucleotide sites in the case of DNA) with n columns, bootstrapping generates a new matrix, also with n columns, by resampling from the original matrix with replacement. Some columns from the original matrix won't be sampled at all, some will be sampled once, and some will be sampled multiple times.

Bootstrapping is used to generate many new matrices (typically at least a hundred, but ideally 1000 or more), and maximum likelihood searches are then run on each matrix. This generates a sample of phylogenies. These can be examined in a variety of ways, but the most common is to evaluate the frequency of each branch of the maximum likelihood tree (generated from the original data matrix) in the sample of bootstrap phylogenies. A branch frequency of 100% indicates that a branch is always in the bootstrap replicates.

Note that I am not using the term "significance" when referring to bootstrap support. This is because bootstraps don't have a clear statistical interpretation. It is a scale that varies from 0 to 1, but is not itself a significance. It just indicates how frequently a branch is recovered when the data columns are resampled. This gives a sense of how broad support is for the branch across characters, but is quite complicated in reality. For example, some variation across bootstrap replicates is due to resampling, but sometimes it is just due to the stochastic nature of heuristic maximum likelihood searches.

Many phylogenetic inference programs do not run full independent maximum likelihood searches on each bootstrap replicate. Instead, they borrow information across replicates, such as optimal starting trees, to speed up the process. One widely used example is the ultrafast bootstrap approximation (Hoang et al., 2018).

6.3 Topology tests

Rather than assess the support of a particular focal topology, sometimes you want to assess how significant the difference in support is for specific phylogenies. There are several different topology tests that address these questions. They include the SOWH test (Swofford et al., 1996), the KH test (Kishino and Hasegawa, 1989), the SH test (Shimodaira and Hasegawa, 1999), and the AU test (Shimodaira, 2002). Some of these are implemented in popular maximum likelihood software tools including IQ-TREE (Minh et al., 2020). Their performance differs across conditions and their assumptions differ, so care should be taken in selecting and evaluating these tests (Goldman et al., 2000; Markowski and Susko, 2024).

Chapter 7

Gene trees and species trees

So far we have gone to great lengths to understand the evolution of single genome regions. Implicitly, we have treated the phylogenies of genes as though they were the phylogenies of the species that carry them. This is a common and often useful simplification, but it is often not correct. In this chapter we will examine the distinction between **gene trees** and **species trees**, and the biological processes that can make gene trees disagree with each other and with the species tree.

Gene trees can disagree due to estimation error. Any tree we infer from a finite stretch of sequence is uncertain, and two genes can appear to disagree simply because we have estimated each of them imperfectly. This in part motivated the evaluation methods we already discussed (Chapter 6). But gene trees can also disagree because the histories of different genome regions are genuinely different (Maddison, 1997). This discordance is the expected outcome of ordinary biological processes acting on real populations and real genomes. A gene tree that disagrees with the species tree can be right about its own history, and quite distinct from the history of the species that carry it. While estimation error shrinks as we collect more data per gene, true discordance does not. It is a property of the history itself.

7.1 Gene trees, species trees, and real discordance

Maddison (1997) provided the framing of gene tree discordance that serves as the foundation for this chapter. Picture the species tree not as a set of thin lines but as a set of tubes, each tube a population extended through time that branches as species split. The history of any particular gene is a gene tree that runs inside these tubes. Sometimes a gene tree follows the shape of its tubes and matches the species tree exactly; sometimes it does not.

Several distinct biological processes can lead gene trees to be discordant with each other and the species tree:

- **Incomplete lineage sorting**, in which ancestral variation persists across successive speciation events and sorts among descendants in a way that does not track the order of speciation. This is the domain of the multispecies coalescent.
- **Gene duplication and loss**, in which a genome region is copied within a genome or deleted, so that the history of the region includes events that the species tree does not.
- **Introgression, hybridization, and horizontal transfer**, in which genetic material moves between lineages that have already diverged, so that a region's history is genuinely reticulate rather than tree-like.

7.2 Incomplete lineage sorting

The process behind incomplete lineage sorting is most naturally described backward in time. Let us consider the process on an example gene tree and species tree (Figure 7.1), knowing that what we learn in this context generalizes. Here we follow a gene sampled in three different species, A, B, and C, back through their ancestral populations. Looking backward, gene lineages **coalesce** when they find a common ancestor. Coalescence between genes sampled from different species will always occur before the species split; the question is how long before. The two lineages first have the chance to coalesce along the branch subtending the speciation, which we will call the **focal branch**. But they could coalesce deeper in the tree.

There are three possible outcomes to consider (Figure 7.1):

1. The A and B lineages coalesce along the focal branch (complete lineage sorting). **The gene tree is concordant with the species tree.**
2. The A and B lineages fail to coalesce along the focal branch (incomplete lineage sorting), but in the ancestral population A and B are still the first pair to coalesce. **The gene tree is concordant with the species tree.**
3. The A and B lineages fail to coalesce along the focal branch (incomplete lineage sorting), and instead A or B coalesces first with C. **The gene tree is discordant with the species tree.**

The first two outcomes give a gene tree that matches the species tree; only the third produces discordance. We can calculate the probabilities of these outcomes with the multispecies coalescent model.

The effective population size N_e is the number of diploid individuals in an idealized population. Since diploid individuals each carry two copies of each gene, the total number of gene copies in a population of N_e diploids is $2N_e$. Given a particular gene sampled from the population, the probability that it coalesces with another particular gene in the previous generation is $1/(2N_e)$. From this, we can derive that the expected time back to the most recent common ancestor of any two copies of a gene sampled from the population is $2N_e$ generations. The larger the effective population size, the longer we expect it to take for lineages to coalesce.

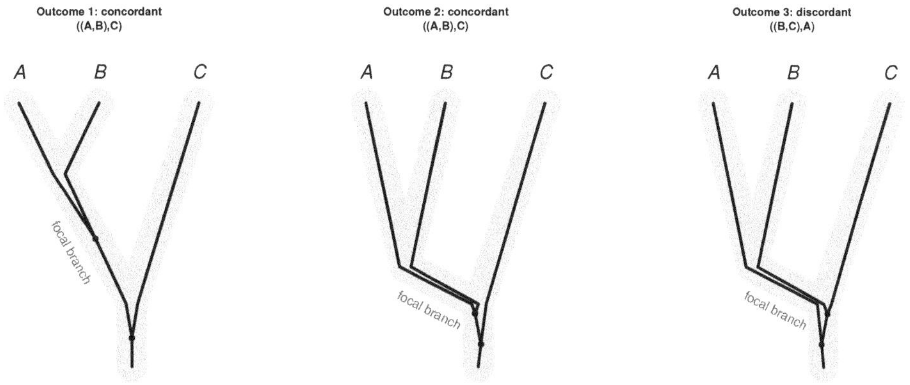

Figure 7.1: The three possible outcomes for a single gene sampled in species A, B, and C, whose species tree is ((A,B),C). Gene lineages (black) evolve within the species tree (gray tubes); the focal branch is the branch ancestral to the A–B speciation, where the A and B lineages first have the chance to coalesce. Outcome 1: the A and B lineages coalesce along the focal branch (complete lineage sorting), giving the concordant gene tree ((A,B),C). Outcome 2: the lineages pass through the focal branch without coalescing (deep coalescence), but A and B are still the first pair to coalesce in the ancestral population, again giving the concordant ((A,B),C). Outcome 3: after the same deep coalescence, B instead coalesces with C first, giving the discordant gene tree ((B,C),A).

Since generation times and N_e can vary widely across species, branch lengths are often normalized to **coalescent units.**

$$\tau = \frac{t}{2N_e} \tag{7.1}$$

A branch length of $\tau = 1$ coalescent unit is the expected time for two lineages to coalesce within a population.

As in our analyses of sequence evolution, we would like to have probabilities, not just expectations. The probability that two copies of a gene sampled from a population fail to coalesce along a branch of length τ is:

$$P(\text{fail to coalesce}) = e^{-\tau} \tag{7.2}$$

Note that the exponent contains a negative sign, so the probability of failure to coalesce decreases as the branch length increases. Equivalently, the probability of coalescence increases as the branch length increases. This makes sense. The more time two lineages spend together in a population, the more likely they are to coalesce.

Equation (7.2) gives the probability that outcome 2 or outcome 3 occurs. To split that probability between these two outcomes, we need a short combinatorial assessment. When A and B fail to coalesce, all three lineages enter the ancestral population together, and the first coalescence there is equally likely to join any of the three pairs: A with B, A with C, or B with C. Only one of these, A with B, matches the species tree. So, given that A and B failed to coalesce along the focal branch, the gene tree is concordant with probability 1/3 and discordant with probability 2/3.

Multiplying the probability of failing to coalesce by the probability of a discordant resolution gives the overall probability of a discordant gene tree (outcome 3),

$$P(\text{discordant}) = \frac{2}{3}\, e^{-\tau} \tag{7.3}$$

Since the probabilities sum to 1, the probability that the gene tree matches the species tree (outcomes 1 or 2) is one minus this,

$$P(\text{concordant}) = 1 - \frac{2}{3}\, e^{-\tau}. \tag{7.4}$$

Equation (7.4) is plotted in Figure 7.2: the match probability is 1/3 when the focal branch is vanishingly short ($\tau \to 0$), no better than choosing a topology at random, and rises toward one as the branch lengthens.

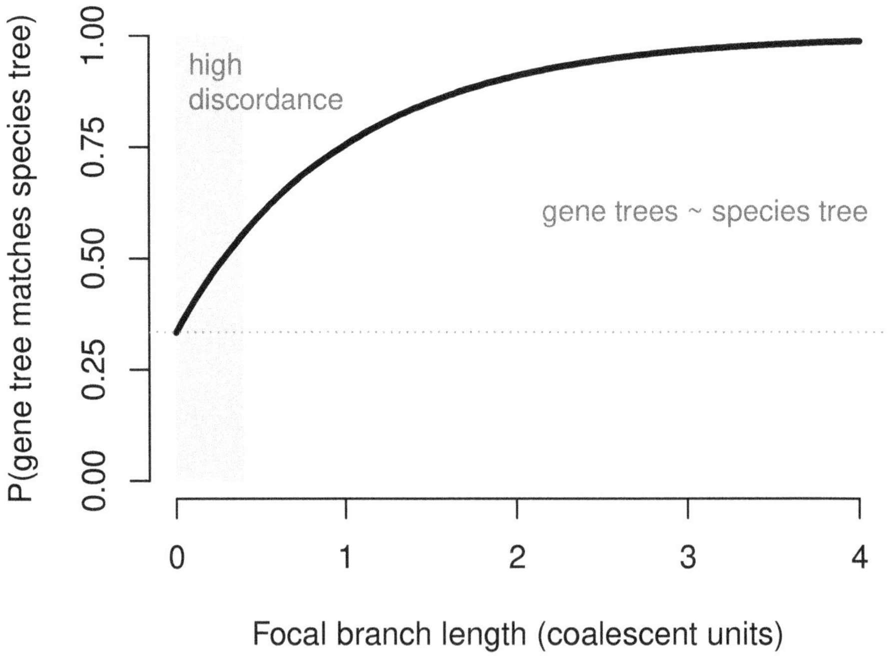

Figure 7.2: The probability that the gene tree matches the species tree rises as the focal branch lengthens (in coalescent units). A short focal branch, as at the base of a rapid radiation, leaves a wide zone of frequent discordance; a long one leaves gene trees essentially concordant.

Although we worked this out for a single focal branch in a three-species tree, the same logic applies to every internal branch of any species tree: the shorter a branch is in coalescent units, the more its gene trees disagree with the species tree. Incomplete lineage sorting is therefore a serious concern wherever internal branches are short—rapid radiations, recent and closely spaced speciation events, and species with large effective population sizes—and negligible where internal branches are long.

Two broad families of coalescent-aware phylogenetic methods, which model and accommodate the multispecies coalescent, are in wide use:

- **Summary, or two-step, methods.** Estimate a gene tree for each region, then estimate the species tree from the distribution of gene trees under the multispecies coalescent. **ASTRAL** is the most widely used method of this kind (Zhang et al., 2018); it is statistically consistent under the coalescent and scales to genome-wide datasets.
- **Co-estimation and site-based methods.** Full-likelihood approaches such as *BEAST co-estimate gene trees and the species tree together in a single Bayesian analysis (Heled and Drummond, 2010), which is powerful but computationally demanding. Site-based methods such as SVDquartets work directly from the sequence patterns under the coalescent without first estimating gene trees (Chifman and Kubatko, 2014).

Edwards (2009) is a good general treatment of species-tree inference under the coalescent.

7.3 Gene duplication and loss

Genome regions are duplicated and lost throughout evolution, and as a result most genes are members of **gene families**: sets of homologous genes within and across genomes that trace back to a history of duplication and differentiation.

This is where the familiar vocabulary of **orthology** and **paralogy** comes in. Two gene copies are orthologs if their most recent common ancestor immediately precedes a speciation event, and paralogs if it precedes a duplication event. The distinction matters in part because our strategy of using genes as proxies for species assumes we are comparing orthologs: copies whose divergence tracks the divergence of species. Compare paralogs by mistake and the gene tree can depart drastically from the species tree.

It is tempting to restrict phylogenetic analyses of species to single-copy orthologs: regions that are present in exactly one copy in every species, with a clean one-to-one correspondence and no duplication or loss to worry about. Though many analyses attempt to approximate this, single-copy orthologs are exceedingly rare in practice. When sampling thousands of genes from even a moderately sized phylogeny of a few dozen species, there may only be a handful of genes that occur as no more than one copy in any species. The biological motivation for excluding genes with any duplication or loss is also less compelling than it may seem at first glance. Duplication and loss are pervasive features of genome evolution; a gene that has remained single copy across a whole clade

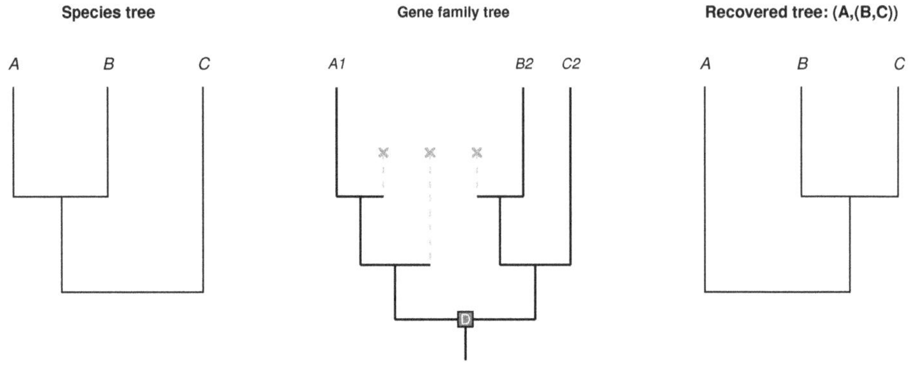

Figure 7.3: How duplication and loss can make single-copy genes mislead. Left: the species tree, ((A,B),C). Middle: a gene family tree in which an early duplication (D) produced two copies; within each copy divergences track speciation, so each copy on its own recapitulates the species tree. Reciprocal loss (dashed gray lineages, marked X) then removes B and C from the first copy and A from the second, leaving exactly one surviving copy in each species: A1, B2, and C2. Right: these survivors look like ordinary single-copy genes, but A1 is a paralog of B2 and C2, so the tree they support is (A,(B,C))—B groups with C, conflicting with the species tree. This hidden paralogy is invisible unless the full gene family history is reconstructed.

for hundreds of millions of years may be unusual in ways (strong constraint, dosage sensitivity) that make it a biased sample of the genome.

A more general response is to treat the labels *ortholog* and *paralog* as summaries of something richer: an explicit, reconstructed history of duplication and loss (Dunn and Munro, 2016). Rather than sorting pairs of genes into two bins (strict orthologs and others), we can infer the full gene family tree and map it onto the species tree, a procedure called reconciliation (Szöllősi et al., 2015; Chen et al., 2000). Reconciliation places each duplication and loss event on the species tree and, in doing so, tells us which copies are orthologs, which are paralogs, and exactly why (Figure 7.3). Orthology and paralogy fall out as consequences of the history rather than serving as the primitive concepts.

Methods that consider gene duplication and loss can reduce inference problems and also let us discard fewer data (since we can relax our gene selection criteria to include genes with more complex histories). Tools that identify gene families and infer their histories, such as OrthoFinder (Emms and Kelly, 2019), make the full complement of gene families available for analysis, and methods such as ASTRAL-Pro estimate the species tree directly from multi-copy gene family trees, accounting for paralogy instead of requiring it to be filtered away first (Zhang et al., 2020). Many studies include genes with more than one copy per species without full reconciliation, using simpler approaches such as pruning some copies from gene trees before species tree inference, based on gene tree structure or branch lengths.

7.4 Reticulation: introgression, hybridization, and horizontal transfer

The coalescent and gene duplication and loss both take the branching of the species tree for granted: lineages, once split, are assumed to stay split, and discordance arises only from how genes sort or duplicate within that fixed history. Reticulation breaks that assumption. Lineages that have already diverged come back into genetic contact and exchange material, so the history of the organisms is no longer a tree but a **network**, with branches that merge as well as split.

Hybridization and **introgression** are two outcomes of the same underlying event—interbreeding between diverged but still cross-fertile lineages—that differ in scale and consequence. Hybridization is the cross itself, and in its most dramatic form it founds a new lineage whose genome is a balanced mosaic of both parents (a homoploid hybrid) or a doubled complement of both parental genomes (an allopolyploid, a common route to new species in plants). Such a hybrid lineage descends from two branches of the species tree, a merger of two lineages. Introgression is the subtler and more common result: when hybrids backcross repeatedly into one of the parental populations, a minority of one lineage's genome leaks into the other while both species persist largely intact. The recipient remains recognizably itself but carries some regions inherited from the donor. Introgression therefore touches only part of the genome, so it is usually detected from genome-wide asymmetries in shared variation rather than from any single locus.

Horizontal, or lateral, **gene transfer** moves genetic material between lineages outside of reproductive processes. Rather than passing from parent to offspring, a gene is taken up directly, carried by a plasmid, a virus, or environmental DNA, and inserted into another genome. Because this requires neither interbreeding nor close relationship, horizontal transfer can move a gene between very distantly related organisms, and it is pervasive among prokaryotes. A transferred gene has a history that leaps across the tree, resembling neither the recipient's close relatives nor a pattern of incomplete lineage sorting.

All three processes produce reticulate histories that no single tree can fully capture, and phylogenetic networks generalize trees to represent them (Solís-Lemus et al., 2017).

7.5 Conclusion

We have examined biological processes that can lead gene trees to depart from each other and from the species tree, methods that model these processes, and approaches to addressing discordance. This is one of the most active areas of phylogenetic research, and there is considerable ongoing development and consideration of the best approaches for particular problems.

Many analyses do not attempt to model discordance at all, and treat all regions as though they share one history. One of the most common such approaches is to concatenate all the gene alignments end to end into one supermatrix and infer a single tree. There are a few reasons this is still a common approach. In practical terms, it often gives the same answer as methods that explicitly accommodate discordance, and it is computationally simpler. In some cases, the biology of the problem may make discordance unlikely. Another concern is that methods that accommodate discordance have many more parameters, and data may be too limited to support the more complex models. In practice, it is common for studies to take multiple approaches and compare the results. Understanding where the methods agree and where they disagree can be informative about the biology of the group and the structure of the problem, and can help identify additional approaches that may be needed to understand the problem.

Real discordance can have multiple causes. Methods have been developed to explicitly model and address each. In reality, the specific processes that cause any particular discordance may be non-identifiable, and the best we can do is to use methods that are robust to multiple causes. The coalescent is a good example: it models one cause of discordance, but it may be able to essentially soak up other causes such as gene duplication and loss. Just because a method detects discordance doesn't mean the discordance is due to the process it models.

We should expect that different processes tend to dominate in different parts of the tree of life, and that the biology of the group and structure of the phylogenetic problem should inform the choice of method. As we have seen, incomplete lineage sorting is more likely on shorter branches and in larger populations. Given that shorter branches tend to be sampled in recent divergences, this is a particular concern at shallower depths in the tree. Incongruence due to gene duplication and loss is most likely when longer times have elapsed along the branches of the species tree. This is a particular concern for deeper phylogenetic problems, such as those regarding relationships from hundreds of millions of years ago.

Gene tree/species tree discordance is one of those phylogenetic patterns that is often treated as a technical nuisance to be overcome in the quest to resolve species relationships. But it is also a rich area of biology that is clearly relevant, and often central, to the larger evolutionary questions that are leading us to infer phylogenies in the first place. Incomplete lineage sorting tells us how population-level processes give rise to macroevolutionary patterns. Gene duplication and loss is central to the evolution of genome novelty and complexity. Introgression and hybridization are central to understanding the evolution of many groups, including humans. As the field develops these processes will be treated less as nuisances to be overcome, and more as important biological phenomena to be sought out and understood.

Chapter 8

Bayesian phylogenetics

Let's take stock of what we have covered so far. We introduced models, and how to simulate data with them. We then explored how to use models to calculate the likelihood—the probability of the observed data given the topology, branch lengths, model parameters, and model. We used likelihood as an optimality criterion in heuristic searches to find the maximum likelihood (ML) phylogeny. We introduced the methods to calculate the frequency of branches on a focal topology, like the ML topology, in a sample of phylogenies. We showed how to generate a sample of phylogenies by running ML searches on bootstrapped matrices, and used this to calculate bootstrap support for each branch in the ML phylogeny.

ML and bootstraps are widely used and are a critical foundation for many phylogenetic analyses. There are, however, a few things about analysis frameworks based on optimality criteria, ML, and bootstraps that are not ideal:

- Whenever we apply an optimality criterion, such as ML or parsimony, to identify the "best" phylogeny, that doesn't tell us anything about how much better this optimal phylogeny is than other hypotheses. If it is the best phylogeny by far, then this single phylogeny tells us a lot about the phylogenetic information in our analysis. If, however, there are multiple phylogenies that are almost as good as the best one, as is often the case, then we would need to know about these others as well to have a good understanding of what hypotheses are consistent with our data.
- Likelihood is the probability of the data given the phylogenetic hypothesis. The probability of the data under the ML hypothesis will be exceptionally low, often far less than one in a million, since the ML hypothesis could generate many other datasets as well. This is because we are evaluating the probability of the data given the hypothesis, but often what we really want to know is the probability of the hypothesis given the data. The two are quite different, but are related. It would be nice to be more explicit about that relationship.

- Bootstraps are a convenient way to generate a sample of phylogenies, but their statistical interpretation is not clear. The frequencies of bootstrap supports don't correspond in a clear way to probabilities of our hypothesis, but instead tell us something about how frequently features of the phylogeny are supported when we resample the data. In practice it turns out this is a good proxy for phylogenetic support, but it would be nice to have support values that have a more explicit statistical interpretation.

All analyses include a variety of tradeoffs, and the issues listed above are often mitigated by the multiple upsides of an ML bootstrap analysis framework. It also greatly helps that there are decades of experience with ML analyses that help contextualize and interpret ML bootstrap results. This work has produced ML methods and software packages that are highly optimized for computational efficiency, enabling the routine application of these approaches to very large datasets.

Another analysis framework, Bayesian statistics, directly addresses all of the issues listed above (Yang and Rannala, 1997; Holder and Lewis, 2003; Huelsenbeck and Ronquist, 2001; Larget and Simon, 1999; Chen et al., 2014). Rather than focus on a single "best" topology, it provides a set of hypotheses consistent with the data. The frequencies of these hypotheses in the sample are the expected probabilities of the hypotheses given the data. These frequencies have a clear probabilistic interpretation, unlike the bootstrap support values. Bayesian statistics and likelihood have deep mathematical connections, so we will be able to build directly on the foundation established in previous chapters.

8.1 Bayesian statistics

Bayesian statistics is much older than many other domains of statistics, but has recently seen a surge in use as computational and methodological advances allow it to be applied more effectively to a wider range of problems. The basic intuition is that our understanding of the world is not drawn from the data alone. Instead, the data are used to update our prior understanding of the world. This updated understanding is referred to as the posterior. There are many excellent introductions to Bayesian statistics, so I will not provide a full derivation here. If you would like more of a background, please consult Appendix A.2.

Bayes' theorem establishes the following relationships:

$$P(H|D) = \frac{P(D|H)P(H)}{P(D)} \tag{8.1}$$

$P(H|D)$, read as "the probability of the hypothesis given the data", is the posterior probability of a given hypothesis. It takes into account prior expectations and new insight from the data. This posterior probability is what we are trying to estimate. We have already seen $P(D|H)$, read as "the probability of the data given the hypothesis". This is the likelihood, and from previous chapters we know how to calculate it. $P(H)$ is the prior probability of the hypothesis, our world view before we collected data. $P(D)$ is the probability of the data averaged over all hypotheses, each weighted by its prior probability. It is not a prior belief about the data in the way that $P(H)$ is a prior belief about the hypothesis; it is the normalizing constant that ensures the posterior sums to one across all hypotheses.

Let's plug in a few numbers to get some intuition for the behavior of Bayes' theorem. First, consider the case where $P(H) = 0$, *i.e.*, you assign a prior probability of 0 to the hypothesis. The posterior probability, $P(H|D)$, will then also be 0. This shows that if you believe ahead of collecting data that the hypothesis is absolutely impossible, no amount of data can change your mind and your updated hypothesis, the posterior, will also be 0.

Second, let's consider the case where the data are just as likely under the hypothesis as they are under other hypotheses. This would be like having a medical test for a specific condition where the test was so bad that the data (test result) didn't depend in any way on the hypothesis (the presence of the condition). For example, the test returns a positive 10% of the time regardless of whether you have the condition or not. In that situation, $P(D|H) = P(D)$. Since $P(D|H)$ is in the numerator and $P(D)$ is in the denominator, they cancel out. That leaves $P(H|D) = P(H)$—the posterior probability of the hypothesis is the same as the prior on the hypothesis. The data didn't change our understanding of the hypothesis at all. This is the behavior we want when the data have no information relevant to the hypothesis.

Third, consider a case where the data are perfect indicators of the hypothesis. You get a specific pattern in the data whenever, and only when, the hypothesis is true. Because the hypothesis and data are perfectly linked, $P(D) = P(H)$, and the priors cancel out. In this case, $P(H|D) = P(D|H)$, *i.e.*, the posterior becomes equal to the likelihood. This never happens in phylogenetic inference, since any given phylogeny could always generate multiple sequences at the tips. But it is generally the case that as the data become more informative, the priors have less impact on the posterior.

8.2 Bayesian phylogenetic inference

To calculate the posterior probability of a phylogenetic hypothesis, we need to address all the terms on the right side of Equation (8.1). We already know how to calculate $P(D|H)$, the likelihood. What about $P(H)$ and $P(D)$? There are a variety of practical approaches to $P(H)$ that have been shown to work well in the context of phylogenetics. These include a uniform prior on topologies, under which all distinct topologies are assigned equal prior probability. Branch length priors are often modeled with an exponential distribution.

What, then, about $P(D)$? Calculating it requires integrating the probability of generating these particular character data (*e.g.*, nucleotide sequences observed at the tips) across all possible topologies and branch lengths, weighted by their priors. That would be prohibitively computationally expensive to actually do. So we won't.

Instead, we will forego calculating $P(D)$ by approximating the posterior with Markov Chain Monte Carlo (MCMC) sampling (Metropolis et al., 1953; Hastings, 1970). MCMC is a widely used method to approximate probability distributions that are too complex to calculate analytically. MCMC is implemented by proposing a series of hypotheses that are either rejected or accepted based on a specially formulated acceptance ratio R and criteria for evaluating it, such that the accepted hypotheses form a sample that is drawn from the distribution of interest. In our case, that distribution of interest is the posterior distribution.

Consider the current hypothesis to be H, and the newly proposed hypothesis H^*. We calculate this acceptance ratio as the ratio of the posterior probability of H^* to the posterior probability of H:

$$R = \frac{P(D|H^*)P(H^*)}{P(D)} \frac{P(D)}{P(D|H)P(H)} = \frac{P(D|H^*)P(H^*)}{P(D|H)P(H)} \tag{8.2}$$

Because $P(D)$ doesn't depend on the hypothesis under consideration, it is the same in both posteriors. Because we are considering a ratio of posteriors, it cancels out. We don't need to calculate it to derive R.

This form assumes the proposal mechanism is symmetric—that proposing H^* from H is as probable as the reverse. When it is not, an additional proposal ratio corrects for the asymmetry; we omit it here for simplicity.

The "Markov Chain" in MCMC alludes to the fact that MCMC is a series of repeated events that depend only on the previous step. Each repeated cycle of events, also known as a generation, proceeds as follows:

1. We have hypothesis H.
2. We propose a new hypothesis H^* by modifying H.
3. We calculate R according to Equation (8.2).
4. If $R \geq 1$, we accept H^*. Otherwise, we accept H^* with probability R, and retain H if it is not accepted.
5. The result of the step above is added to the posterior sample, and becomes H for the next iteration of the cycle.

MCMC produces a sample of model parameters and topologies with branch lengths. This sample is an approximation of the posterior distribution of these entities. We can summarize the topologies in this posterior sample in the same way we did for bootstraps, with branch frequencies. Unlike bootstraps, though, the frequency of a branch in this distribution has a clear statistical interpretation. It is an approximation of the posterior probability of that branch, *i.e.*, the probability of the branch given the data and our priors. The branch lengths and model parameters form continuous probability distributions. We can summarize these in a variety of ways, for example by taking the mean for each.

8.3 Bayesian phylogenetic inference in practice

There are a variety of practical considerations to implementing a Bayesian phylogenetic analysis with MCMC. These analyses are carried out with widely used software packages such as MrBayes (Huelsenbeck and Ronquist, 2001; Ronquist et al., 2012) and BEAST (Drummond and Rambaut, 2007). The practical considerations include:

- Selecting the appropriate model. Many of the same considerations we reviewed in the context of likelihood apply here.
- Selecting appropriate priors on topology, branch length, and model parameters.
- Devising an appropriate hypothesis proposal mechanism. The scale of the proposals does not change the distribution the chain is sampling from—with a correct Metropolis-Hastings implementation the chain converges to the same posterior regardless of proposal scale. What the proposal scale affects is efficiency: how quickly the chain explores the posterior, and therefore how well a run of finite length approximates it. If the proposed hypothesis H^* is usually far from H, most proposals land in regions of low posterior probability and are rejected, so the chain stays in place for long stretches and moves slowly. If the proposals are instead very small, most are accepted but each step barely moves, so the chain drifts slowly and successive samples are highly autocorrelated. Either extreme mixes poorly: successive samples carry little new information, the effective sample size is low, and a run of a given length gives a noisier, less reliable approximation of the posterior. In the worst case a finite run may not have had time to move between well-separated regions of high probability at all, so it misses part of the posterior even though the target distribution it would eventually sample is correct. Good proposals strike a balance, and there are several approaches to tuning them, including searches with multiple MCMC chains that take different step sizes.
- The very early samples in the chain are not a good estimate of the posterior distribution, because the initial H is not necessarily close to a peak in the posterior. The initial samples before MCMC settles into a stable distribution are therefore discarded as part of a "burn-in" phase.
- Knowing when enough generations have been sampled to adequately approximate the posterior distribution. This is often evaluated by running multiple independent MCMC analyses and stopping them if and when they have converged on similar distributions.

The ability to specify a prior can be a great advantage of Bayesian analyses. A prior is an explicit, principled way to incorporate prior knowledge into a new analysis. For example, if you are very confident that a particular clade exists based on external evidence, you could design a topology prior that favors topologies that include that clade over topologies that don't. If you have external information about model parameters, you could set a prior that constrains these parameters. These are cases of what are called informative priors—they are nonuniform, and intentionally constrain results based on external information. In many cases, though, investigators use relatively diffuse, or vague, priors that spread probability broadly across hypotheses rather than concentrating it. If you are ever concerned about the impact of your priors on your posteriors, you can run your analysis without any data—the posterior will then be determined entirely by the prior. This gives you a good understanding of the impact of your prior on the posterior, and by comparing this empty analysis to an analysis with data you can get a good sense of the impact of your data on the posterior.

Bayesian phylogenetic software stores samples of the run at regular intervals, for example every hundred generations (where a generation is a single MCMC proposal). Typically, there are two output files, a tree file and a trace file. The tree file has one line per sample, each with a tree in Newick format. The trace file also has one line per sample. There are multiple columns, each with a model parameter or run statistic. It is called a trace file because you can plot each of these columns to trace the progress of the MCMC run.

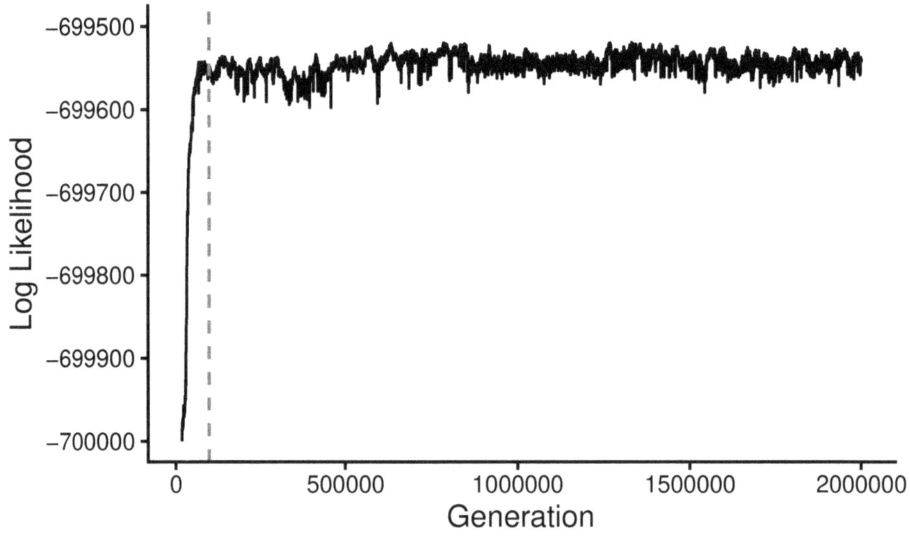

Figure 8.1: Plot of log likelihood trace for a Bayesian analysis. A burn-in of 100,000 generations is indicated with a dashed vertical line.

Investigators should at minimum examine the trace of log likelihood (Figure 8.1). There are a few things to look for. First, an appropriate burn-in should be identified. This is the number of generations at the start of the run that are discarded as the MCMC chain settles into the sampling process. During the burn-in, the log likelihood rapidly climbs as the chain moves toward higher likelihood regions of tree and model space. In the example provided here, the chain appears to have moved past this initial climb after roughly 100,000 generations. Second, the investigator should examine the post-burn-in sampling process to make sure it looks well mixed. It should appear noisy, as it does here, indicating that trees and model parameters with a range of likelihoods are sampled, and it should not show sudden major jumps or longer-term trends, which can indicate inadequate sampling.

A noisy, trend-free log likelihood trace is necessary but not sufficient evidence that a run is sampling the posterior well. A chain can look perfectly stationary in log likelihood while remaining stuck in one region of tree space, or while individual parameters remain poorly sampled, and the log likelihood alone cannot reveal this. Nor does choosing a burn-in establish convergence: discarding the initial samples only removes the early transient before the chain reached its current region. If a chain has not mixed or converged, no choice of burn-in can repair it.

Other diagnostics are available to assess whether a run has adequately sampled the posterior. These include examining the traces of individual model parameters and calculating the effective sample size (ESS) for each parameter. The ESS estimates how many effectively independent samples the autocorrelated chain provides for a given parameter; a low ESS (a common rule of thumb flags values below a few hundred) signals high autocorrelation and poor mixing, and calls for a longer run or better proposals.

Once the trace file has been examined, the investigator turns to the tree file. All pre-burn-in trees are discarded. You then pick a focal tree, which could be the maximum likelihood tree or consensus of the post-burn-in Bayesian trees. The frequency of each edge in the focal tree is then calculated from the set of post-burn-in trees and recorded as the posterior probability of that edge. These are then plotted onto the tree, often along with the maximum likelihood bootstrap scores.

8.4 Resources

- A video introduction to Bayes' theorem: https://www.youtube.com/watch?v=HZGCoVF3YvM
- The excellent MCMC robots by Paul Lewis: https://plewis.github.io/applets/mcmc-robot/

Chapter 9

Additional character types

So far we have considered a single type of character data—DNA sequences. But there are many other types of characters that we would like to measure and analyze on phylogenies. These include morphology, protein sequences, protein structure, gene expression, physiological traits, and environmental tolerances. Different types of character data need to be handled in different ways. In particular, we need to be able to articulate explicit models for how each type of data evolves.

We can group character types based on shared features. Some data, such as the nucleotides of DNA, are discrete unordered values from a finite set $\{A, C, G, T\}$ where the elements have no inherent ordering (*e.g.*, C is not greater than A, or less than T). Others have continuous values, such as mass or length, that fall in a particular order on the real number line (2.5 is less than 3.2). Still others, such as the number of bristles on a leg segment, have countable values represented with integers. These values are both discrete (there are no integers between 3 and 4) and ordered (3 is less than 4). If we consider these different character types, it is clear not only that we measure and record them differently, but that the measurements themselves have different properties requiring different analysis methods. We could not, for example, construct a matrix like Equation (3.5) with rows and columns for every possible length in millimeters for flower petals.

Rather than approach each new character type in an *ad hoc* way, it is better to examine their general properties and explicitly consider how each character should be encoded and modeled. Specifying the character types is a critical aspect of how we articulate our ontological perspective (*i.e.*, what organismal attributes exist, which are worth considering for the question at hand, and what the relation between them is). The character type your data correspond to is an intrinsic property that must be correctly identified. The criteria for making this evaluation and understanding its implications are part of measurement theory (Houle et al., 2011). This field sits at the intersection of math, statistics, and philosophy. It considers the relationships between measurements and the reality they represent, and clarifies what information the measurements contain. It

Table 9.1: Scale types, modified from Houle (2011).

Scale type	Domain	Measurement type	Meaningful comparisons	Biological examples
Nominal	Any set of symbols	Discrete	Equivalence	Species, genes
Ordinal	Ordered symbols	Discrete	Order	Social dominance
Interval	Real numbers	Continuous	Order, differences	Dates, Malthusian fitness, relative temperature (arbitrary 0, e.g., Celsius and Fahrenheit)
Log-interval	Positive real numbers	Continuous	Order, ratios	Body size
Difference	Real numbers	Continuous	Order, differences	Log-transformed ratio-scale variables
Ratio	Positive real numbers	Continuous	Order, ratios, differences	Length, mass, duration, absolute temperature (e.g., Kelvin)
Absolute	Defined	Continuous	Any	Probability

examines which mathematical operations we can perform with them while preserving their meaning, and reveals what actual transformations those operations correspond to. With a name like "measurement theory", you might assume that it is a dusty and boring annoyance that someone else needs to worry about, but it is actually a fascinating and grounding framework for understanding many of the central aspects of what we do in science.

Because biological measurement developed pragmatically rather than from measurement theory, there are some differences in the nomenclature. What phylogenetic biologists call "character type" is referred to in measurement theory, and many other fields of science, as "scale type" (Table 9.1). Scale types vary in several ways. The Domain indicates the possible values. Phylogenetic methods differ most based on whether this domain is discrete or continuous, reflected here in the Measurement type column. Meaningful comparisons indicates comparisons that can be made between measurements of each scale type. Our comparisons should be constrained to respect the physical realities that our measurements represent, and measurement theory gives us formal tools to evaluate this. We should not, for example, compute the numerical difference between ordinal states, like social dominance ranks, because the actual physical or evolutionary distances between those ranks are unknown and unequal.

There are many types of organism measurements, and therefore state spaces and character types, that are addressed in a phylogenetic context. Here we consider some of the more frequently applied character types, *i.e.*, scale types. Different scale types require different models of evolution.

9.1 Discrete character types

9.1.1 Nominal scale types

9.1.1.1 DNA nucleotides

Measurements of DNA sequences have 4 possible states, corresponding to each of the 4 nucleotides—$\{A, C, G, T\}$. DNA data are discrete and unordered. Nucleotides are discrete because they have a set of distinct and separate states. They are unordered because changes don't have to occur in a specific order; any state can change to any other state directly. In measurement theory terms, discrete unordered character types correspond to a nominal scale type.

9.1.1.2 Amino acids

Protein sequences are handled very similarly to DNA sequences, the character states just correspond to amino acids rather than to DNA nucleotides. They are discrete and unordered, and therefore on a nominal scale type. There are 20 possible states instead of 4, so the state space is larger. In principle this implies many more parameters, because the size of the rate matrix grows with the square of the number of states: a fully parameterized model of amino acid evolution has far more exchangeability parameters than a nucleotide model. In practice, however, these are rarely all estimated from the data at hand. Most protein analyses instead use empirical exchangeability matrices, such as JTT, WAG, or LG, whose values were estimated once from large databases of protein alignments and are then held fixed. As a result, the number of parameters actually estimated in a typical protein analysis can be comparable to, or even fewer than, in a nucleotide model, even though the underlying model describes a much larger state space.

There are a few reasons why protein sequences are often considered in phylogenetic analyses rather than the DNA sequences that encode them. One is that questions about protein evolution are best addressed with models that directly describe protein evolution. Another is that synonymous changes in protein-coding DNA sequences quickly saturate for more distant evolutionary comparisons. This means that much of the variation in DNA sequence has little information about phylogenetic relationships. Protein data can be more tractable to work with in this situation.

9.1.1.3 Codons

Since there are 4 possible DNA nucleotides and codons are 3 nucleotides long, there are $4^3 = 64$ possible codons. Each one of these codons corresponds to a specific amino acid or stop codon. In some cases, it is most interesting to consider each of the 64 codons as a discrete character state. The models then have matrices with dimensions of 64×64 (as opposed to 4 for nucleotides and 20 for amino acids).

9.1.1.4 Morphology

Direct analogs of the DNA sequence evolution models are often applied to discrete unordered morphological traits, such as the presence or absence of limbs (Harmon, 2018, chapter 7). The most widely used of these is the Mk model (Lewis, 2001), a k-state generalization of the Jukes-Cantor model in which all changes among k states occur at equal rates.

9.1.2 Ordinal scale types

Ordinal scale types include any kind of ranking, such as position in a social hierarchy (Houle et al., 2011, Table 1). They have an ordering, *i.e.*, some values are larger than others. But there is no statement about the distance between the values. In a pecking order for chickens, 1 is dominant over 2 and 2 over 3, but that doesn't indicate that there is a similar difference between 1 and 2 as there is between 2 and 3.

9.1.3 Discrete ordered characters

Discrete ordered characters are countable traits, such as the number of digits on a forelimb or the number of bristles on an arthropod appendage. Like ordinal characters they have an order, but unlike them the spacing between adjacent values is uniform and meaningful—the difference between 3 and 4 digits is the same as between 4 and 5. This uniform spacing is the defining feature of an interval scale (Houle et al., 2011, Table 1); because counts also have a true zero, however, they are strictly a ratio (or absolute) scale rather than a true interval scale. For modeling their evolution, what matters is that they are discrete and ordered with uniform steps between adjacent states.

Models for the evolution of discrete ordered characters can be described with the same language we used for nominal scale types. The rates for changes between non-adjacent values are just set to zero. 5, for example, will have a nonzero rate of change to 6 and 4 and a rate of zero to all other values. In this way, the rate matrix disallows instantaneous changes that skip intermediate values. For example, to evolve from a forelimb with 5 digits to one with 3 digits, the model requires that the character pass through an intermediate state of 4 digits.

Such a rate matrix describing the changes between 0-6 digits would have this form, if the rates were the same between all states:

$$\mathbf{Q} = \begin{pmatrix} -\mu & \mu & 0 & 0 & 0 & 0 & 0 \\ \mu & -2\mu & \mu & 0 & 0 & 0 & 0 \\ 0 & \mu & -2\mu & \mu & 0 & 0 & 0 \\ 0 & 0 & \mu & -2\mu & \mu & 0 & 0 \\ 0 & 0 & 0 & \mu & -2\mu & \mu & 0 \\ 0 & 0 & 0 & 0 & \mu & -2\mu & \mu \\ 0 & 0 & 0 & 0 & 0 & \mu & -\mu \end{pmatrix} \tag{9.1}$$

9.2 Continuous data

Many characters, such as body mass, limb length, protein abundance, maximum swimming speed, and metabolic rate, can take values across a continuous range of real numbers. In phylogenetics, these traits are typically grouped under the umbrella of continuous character data, since values can vary smoothly and intermediate values are possible between any two observations.

Measurement theory distinguishes several different scale types (for example, interval and ratio scales) that may all have continuous values. In phylogenetic comparative analyses, these distinctions are usually ignored, and the traits are analyzed using a common set of evolutionary models. Sometimes, ignoring these distinctions has little practical impact, but sometimes it has major consequences.

9.2.1 Brownian motion

The most widely used model for continuous trait evolution on phylogenies is Brownian motion (BM) (Felsenstein, 1973, 1985b). In this model, trait evolution is treated as a stochastic diffusion process along the branches of the phylogenetic tree. Trait values change through time by accumulating small random fluctuations. This can be conceptualized as infinitesimally small random steps along the number line, sometimes increasing the value and sometimes decreasing it.

Under BM, the expected change in the trait value is zero. This is not because there is no change under Brownian motion. It is because increases are just as common as decreases. If you run the same BM simulation many times and average the results, the average change across all of them will be zero. Each one, however, can be quite different from the original value and from each other (Figure 9.1). This is because the variance of the trait value increases linearly with time. The longer two lineages evolve independently, or the faster the change, the more different their trait values are expected to become.

Mathematically, Brownian motion can be written as

$$dX(t) \sim \mathcal{N}(0, \sigma^2 dt)$$

where $X(t)$ is the trait value at time t, and σ^2 is the diffusion rate, describing the amount of variance accumulated per unit of evolutionary time.

In phylogenetic applications of the Brownian motion model there are two primary parameters. One is the initial state, X_0, which is the initial trait value. The other is the diffusion rate, σ^2, which determines how quickly trait variance accumulates through time. Along a branch of length t, the trait value at the descendant node is modeled as

$$X_{\text{descendant}} \sim \mathcal{N}(X_{\text{ancestor}}, \sigma^2 t)$$

This means that the expected trait value at the end of the branch is equal to the value at the beginning of the branch, while the variance around that expectation grows in proportion to the branch length.

One reason BM is widely used is that it leads to a convenient statistical formulation. The vector of trait values observed at the tips of a phylogeny follows a multivariate normal distribution whose covariance structure is determined by the shared branch lengths of the tree. This property allows efficient likelihood-based and Bayesian inference.

The BM model is primarily used because it provides a mathematically tractable approximation to trait evolution. However, many biological traits do not actually evolve according to the assumptions of BM and violate the fundamental nature of the measurements being analyzed. For example, BM allows trait values to take any real number, including negative values. Many biological measurements, such as body mass or metabolic rate, are strictly positive. Sometimes these violations have little impact on inference, especially when the observed trait variation is far from biologically impossible values. In other cases, however, they can lead to misleading conclusions if the model poorly reflects the underlying evolutionary processes.

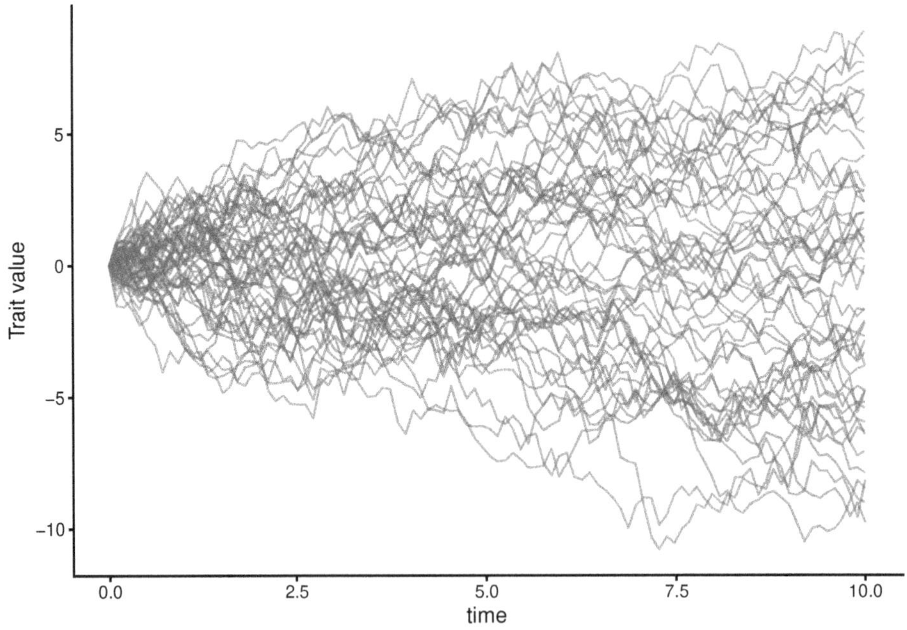

Figure 9.1: Multiple Brownian motion trajectories.

9.2.2 Ornstein–Uhlenbeck models

Brownian motion assumes that there is no preferred trait value or directional trend in evolution. In reality, traits are often subject to constraints, such as stabilizing selection or metabolic constraints, that lead values to stay near a particular optimum. The Ornstein–Uhlenbeck (OU) model is a common variation of Brownian motion that can accommodate these constraints (Hansen, 1997; Butler and King, 2004).

In the OU process, trait evolution is still driven by random fluctuations, but there is also a deterministic tendency for the trait value to move toward an optimum. The model can be written as

$$dX(t) = \alpha(\theta - X(t))dt + \sigma dW(t)$$

where $X(t)$ is the trait value at time t, θ is the value toward which the trait is drawn (often interpreted as an optimal trait value), α describes the strength of attraction toward that value, and σ controls the magnitude of random fluctuations.

The OU model therefore includes both stochastic variation and a restoring force that pulls the trait toward an optimal value. When α is large, trait values tend to remain close to the optimum. When α is small, the restoring force is weak and trait values can wander more widely.

The Brownian motion model can be understood as a special case of the OU model. If the strength of attraction toward the optimum approaches zero ($\alpha = 0$), the restoring force disappears and the process reduces to pure Brownian motion. In this sense, OU models generalize Brownian motion by allowing processes in which trait values are pulled toward a central value rather than wandering freely.

It is important to be careful about the biological interpretation of these parameters. The OU model describes a statistical tendency for a trait to revert toward a central value; it does not, on its own, reveal why. Identifying θ with an adaptive optimum, and a nonzero α with stabilizing selection, is a biological interpretation that depends on the assumptions of the model (Hansen, 1997; Butler and King, 2004). A trait can appear to evolve under an OU process for reasons unrelated to adaptation, including bounds on the range of possible values, measurement error, or model misspecification. A good fit to an OU model is therefore consistent with stabilizing selection toward an optimum, but does not by itself demonstrate it, and inferring the underlying evolutionary process from the fit of a phylogenetic model alone is notoriously difficult (Uyeda et al., 2018). Distinguishing genuine adaptation from these alternatives generally requires additional evidence, such as *a priori* hypotheses about which lineages are expected to share an optimum.

Chapter 10

Phylogenies and time

Phylogenies consider the history, pattern, and process of evolution through time, so time is often a critical feature of phylogenetic analyses. The specific considerations of time vary across analyses and studies, depending on the data available and the questions asked.

It is helpful to step back and think about how the elements of a phylogeny correspond to time. We will here consider rooted trees, where the direction of time is specified and runs from the root to the tips. Each node occurs at a specific point in time, even if the specific time is unknown. We will call this point in time the node age. There is often confusion about the description of relative time on phylogenies. I will describe the magnitude of age here as time before the present. A node with minimum age is closer to the present (further forward in time and closer to the tips), and a node with maximum age is further from the present (further back in time). A node is treated as a singular divergence event, and has no duration. This of course is an approximation of the actual biology of divergence, which can take place over a period of time as populations become increasingly isolated.

Each edge (branch) connects two nodes. In a rooted tree, we refer to the node closer to the root (or that is the root) as the parent node, and the node closer to the tips (or that is a tip) as the child node. The starting and ending times of the edge are set by the ages of the parent and child nodes. The duration of the edge is the difference in the ages of these nodes.

10.1 Measurements of time on trees

As discussed in Section 2.5, branch length can mean different things. It is up to the investigator to specify a branch length and clearly communicate what it means in the tree at hand. The three usual approaches are a cladogram (a tree in which branch lengths have no meaning), a phylogram (branch lengths are the expected amount of evolutionary change in the traits used to infer the phylogeny), and a chronogram (branch lengths are in units of time). These tree types differ in what we can say about time.

10.1.1 Cladograms

Because branch lengths in a cladogram have no meaning, we cannot make absolute statements about time in a cladogram. This doesn't mean, though, that we can't say anything about time—we can still make some ordinal statements about the relative ages of nodes.

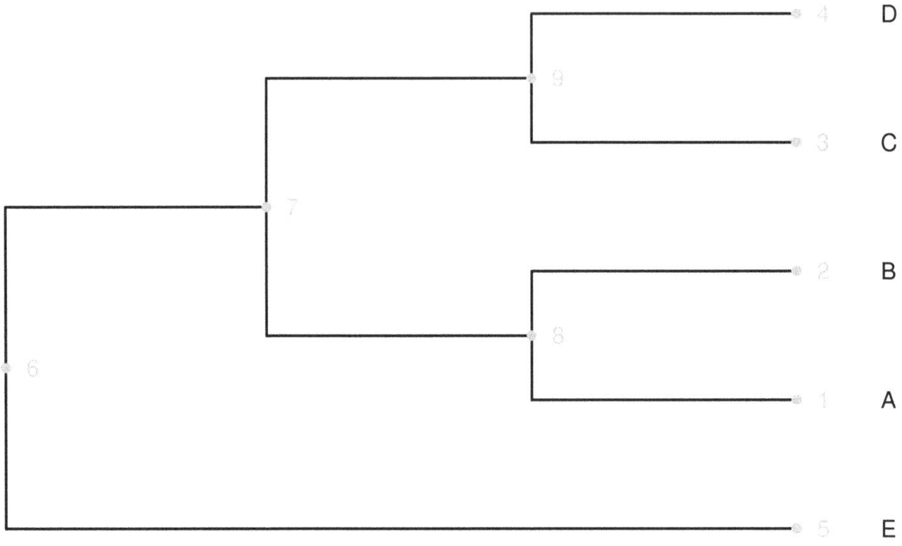

Figure 10.1: A cladogram. Nodes and node numbers are gray, and branches are black.

Take a look at the cladogram in Figure 10.1. Consider the gray node numbers. The terminal nodes are numbered 1-5, and 6-9 are internal nodes. Of those, the root is node 6. Because the tree is rooted, we know that time proceeds from the root to the tips. If you consider two nodes, where one is descended from the other, then the node closer to the root is older. There are a variety of statements we could make based on this simple relationship, including:

- Node 6 is older than all other nodes in the phylogeny. This is tautological, since the root is by definition the oldest node.
- Node 7 is older than node 9. This is because 9 is descended from 7, and 7 is closer to the root.

We can't make any relative assertions, though, about the ages of nodes that aren't descended from each other. For example, we have no idea if node 8 or node 9 is older.

In practice, these relative statements of age are often quite useful. For example, if Clade A is nested within Clade B, then we know that Clade A is younger than Clade B even if we don't know the ages of any nodes. This may seem trivial, but it is important information that is implicit in many discussions of phylogenies.

10.1.2 Phylograms

Strictly speaking, the only assertions we can make about time on phylograms are those that we can also make on cladograms. This is because we can't be certain of the relationships between time and branch length in the phylogram. If the rate of evolution changes in different regions of the tree, then branches of the same length would not represent time intervals of the same length. Phylograms therefore have the same information about time as cladograms do.

10.1.3 Chronograms

By definition, in a chronogram branch lengths are in units of time. The age of each node is specified. This has a couple of implications for what we can say about time. We can make ordinal statements, as we did for cladograms and phylograms. But we can also make absolute statements about the interval of time that has elapsed between two nodes. We can also make statements about the relative ages of nodes that are not descended from each other.

10.2 Time calibration

The process of creating a chronogram is referred to as time calibration. In essence, external information about the ages of some nodes is used to constrain the ages of other nodes.

If the rate of evolution were uniform and did not vary in different lineages, then the branch lengths on the phylogram would be proportional to the elapsed time. We could convert from the phylogram to the chronogram just by dividing branch lengths by the rate of change. For example, if the branch lengths are in number of expected DNA substitutions, and the rate of evolutionary change is always 2 substitutions per million years, then dividing each branch length by 2 would give a chronogram where the units of branch length are millions of years. This is called clock-like evolution—changes in characters are like regular ticks on a clock (Zuckerkandl and Pauling, 1965). No real data evolve in a perfectly clock-like way, so time calibration of phylogenies instead relies on relaxed clock models that allow local variations in evolutionary rate (Sanderson, 1997; Drummond et al., 2006).

If all the tip nodes of a phylogeny are the same distance from the root, the tree is referred to as being ultrametric. If all the tips on a chronogram are sampled at the same time, then the chronogram is expected to be ultrametric because the same amount of time has elapsed from the root to each tip. An ultrametric tree is not necessarily a chronogram, though. Some of the internal nodes may have ages that violate the expectation that branch lengths are proportional to time, or a phylogram could be constrained to be ultrametric without calibrating any of the branch lengths according to time.

In the early days of the field, phylogenetic inference was done separately from time calibration. A phylogram would first be inferred. The phylogram would then be modified into a chronogram. This would involve stretching and shrinking branches to make the tree ultrametric and to constrain some internal nodes to fit fossil calibration dates. This modification of branch lengths was not made with consideration of the underlying trait data.

Subsequent approaches instead unify phylogenetic inference and time calibration. Using external information, such as fossils and the sampling times of tips, branch lengths are inferred in units of time under relaxed clock models that allow rates of evolution to vary across the tree to accommodate these constraints. Because the external calibrations allow rates and time to be inferred independently (unlike an unconstrained phylogram, where they are conflated), this time-calibrated inference produces a chronogram.

10.3 The implications of constraining node ages

Clamping the ages of nodes reduces the number of free parameters in our tree. The way to think about this is that the more information we have, the more constrained and specific our view of the world is. Consider first constraining tip ages to make a tree ultrametric. Before we clamp the tip ages, any tip can be any age and all the branches are free to have any length. The ultrametric tree is nested within this set of unconstrained possibilities. The tip ages, and therefore branch lengths, in this unconstrained tree can be selected so that they are ultrametric, but the vast majority of values will lead to trees that are not ultrametric. By clamping some values with added information that some nodes (tips, in this case) are the same age, we now are allowing only a constrained subset of trees and these require fewer parameters to describe.

Consider this same task in terms of branch lengths, rather than node ages. From Section 2.3, we know that the number of branches in a tree is $2n - 2$, where n is the number of tips. This is because each of the n tip nodes has a branch leading to it, and each of the $n-1$ internal nodes, with the exception of the root node, has a branch leading to it. So there are $n - 2$ branches that give rise to internal nodes. This is how we arrive at our $n + n - 2 = 2n - 2$ branches in the phylogeny, each with its own length. In an ultrametric tree, by definition all the tips have the same age. That means that if you know the length of one of the branches leading to a tip node, you can calculate all the others. They have a deterministic relationship and are not free to vary independently. Rather than n branch lengths for the tips, we only have 1 tip branch length that is free to vary and we can calculate all the others so that the tip nodes have the same ages. This leaves us with $1 + n - 2 = n - 1$ branch lengths that we need to estimate independently in our ultrametric tree.

10.3.1 Time calibration in practice

The use of character data and time calibrations to infer time-calibrated phylogenies falls into two broad categories. The first is to use information from fossils (or other historical data) to constrain the ages of internal nodes. A Bayesian approach provides a natural framework for this—node ages can be specified as priors on the tree. For example, if a given fossil is known to fall within a particular clade, then the most recent common ancestor of that clade can be constrained to be at least as old as the fossil and possibly older.

While this can work well in some situations, there are several drawbacks to this method. In practice, it is also necessary to place maximum ages on some of the nodes to keep the time calibration from pushing everything way back in time. Given the incompleteness of the fossil record, this is not as straightforward as constraining the minimum age of a node. If we have a fossil we can assert that the clade that contains it can be no younger than that specimen, though it may be much older. But it is harder to assert a maximum age, given that we just might not have fossils for older organisms that existed in the clade. Applying maximum ages often relies on expertise and additional information, such as knowledge that a given land mass where the organisms are exclusively found did not exist before a particular time. Careful justification of fossil calibrations is essential, and best practices have been formalized to make them explicit and reproducible (Parham et al., 2012).

The model known as the "fossilized birth–death" (FBD) process doesn't use fossils as external constraints on internal nodes; instead it includes fossils right in the tree as their own tips (Heath et al., 2014). Unlike extant tips, which all have the same age (*i.e.*, today), the ages of these fossil tips are then constrained with geological data. This approach provides the ability to also include parameters for fossil preservation and sampling. FBD methods require that the character matrix includes data that can be scored in the fossils. The FBD process has since grown into a broad family of models with applications spanning paleobiology, macroevolution, and epidemiology; Wright et al. (2022) provide an accessible overview.

Chapter 11

Ancestral character state reconstruction

We learned about models of evolution and how to run these models forward in time, from the root to the tips, to simulate the evolution of characters. In the chapters on inference, we used the same machinery to estimate the tree and model parameters that best explain the states we see at the tips. Here we turn to a different product of the same tools—learning about the states of the characters at the internal nodes of the tree. Estimating these states is called ancestral character state reconstruction (Joy et al., 2016; Revell, 2025).

Recall the framing from Chapter 1. A phylogenetic analysis has several components—the tree, the branch lengths, the model, the model parameters, the character states at the tips, and the character states at the ancestors. Different studies clamp some of these, estimate others, and then keep or discard the estimates depending on the question. Ancestral character state reconstruction is the case where the character states at the internal nodes are what we want to estimate and keep. To get them, we usually clamp the tree, the branch lengths, and the observed states at the tips, and then estimate the states at the ancestors.

There are many applications of ancestral character state reconstruction, and learning about long-dead ancestors has been a central goal of evolutionary biology (Pagel, 1999). These methods can seem like a time machine—they can give us a glimpse into the past, indicating the color of a flower, the presence of a limb, or the body size of an organism that lived tens of millions of years ago and left no direct record. It is one of the most evocative things we can do with a phylogeny, and also one of the easiest to misuse. A reconstruction is not an observation but an estimate, only as good as the model and data behind it, and we take up the cautions that come with that at the end of the chapter.

11.1 Reconstructing discrete characters

We already have all of the machinery we need to reconstruct discrete ancestral states under a model of evolution. In the chapters on simulation and inference we described the evolution of a discrete character with a rate matrix $\mathbf{Q}$, and we saw that the probability of moving from one state to another along a branch of length t is given by the matrix exponential $\mathbf{P}(t) = e^{\mathbf{Q}t}$ (Equation (3.7)). The same model that let us simulate data and compute the likelihood of a tree also lets us compute the probability of each state at each internal node.

The simplest Markovian model for a discrete character with k states is the Mk model (Lewis, 2001), a direct generalization of the Jukes-Cantor model to k states in which all changes occur at the same rate. Just as with DNA models, we can relax this assumption to allow different rates for different transitions.

There are two distinct questions we might ask about the ancestral states, and it is important to keep them separate. The first is the *marginal* reconstruction—for a single node, what is the probability of each possible state, averaging over all possible states at every other internal node? The second is the *joint* reconstruction—what is the single most probable assignment of states to *all* internal nodes considered together? The marginal reconstruction is the more commonly reported, because it gives us a probability for each state at each node, and therefore an explicit measure of uncertainty. It is the one we focus on here.

To see how these probabilities are actually computed, recall how we found the likelihood of a tree in the chapter on inference. We listed every possible combination of states at the internal nodes—every possible history that could connect the observed tips—and computed the probability of each one as the product of the per-branch transition probabilities from $\mathbf{P}(t) = e^{\mathbf{Q}t}$ (Equation (4.1)) together with the frequency of the state at the root. Summing over all of those histories gave the likelihood of the observed tip states.

Marginal reconstruction reuses exactly that calculation. To find the probability that one particular internal node is in a given state, we add up the probabilities of just those histories in which that node holds that state, and divide by the sum over all histories. Repeating this for each possible state gives the set of probabilities we draw as the pie at that node. Because every history threads from the root, through the node, and out to the tips, this ratio automatically folds in the states of the descendants below the node, the states elsewhere in the tree, and the branch lengths on every side. That is why a node flanked by short branches leading to tips that agree is reconstructed sharply, while one reached only across long branches is pulled toward an uninformative, near-even split.

Listing every history is only feasible on very small trees. In practice the same sums are obtained efficiently by a two-pass traversal of the tree—a downpass that gathers information from each node's descendants and an uppass that gathers it from the rest of the tree—but the quantity being computed is precisely the ratio just described. The *joint* reconstruction asks a different question of the same histories: rather than summing, it picks out the single assignment of states to all internal nodes at once that has the highest probability.

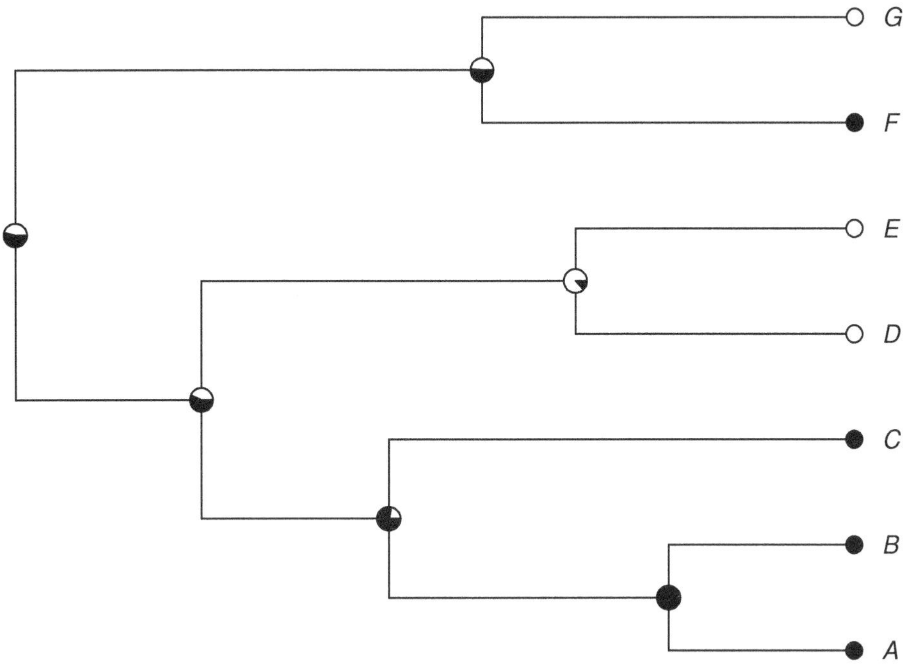

Figure 11.1: Marginal ancestral state reconstruction of a binary character under an equal-rates Mk model. Tips are scored as present (black) or absent (white). Each pie at an internal node shows the estimated probability of each state at that node. Nodes near well-sampled clusters of tips are reconstructed with high confidence, while deeper nodes are more uncertain.

Figure 11.1 shows a reconstruction of this kind for a small binary character. Note that the reconstruction is not a set of hard assignments but a set of probabilities, drawn here as pie charts at each node. Nodes that sit just above a cluster of tips that agree with each other are reconstructed with high confidence. Deeper nodes, where the descendant tips disagree and long branches allow ample opportunity for change, are reconstructed with much more uncertainty. Those uncertain pies are a feature, not a nuisance. They are the model honestly telling us that the data do not strongly favor one state over another.

The reconstruction depends on the model. If we had allowed the rate of gains to differ from the rate of losses, or fit a model in which one state is much harder to leave than another, the reconstructed probabilities at the internal nodes could shift substantially. This is why model selection, discussed in the chapter on evaluating models, matters as much here as it does in phylogenetic inference.

11.2 Reconstructing continuous characters

For continuous characters we reconstruct ancestral states under a continuous model of evolution, most often the Brownian motion model introduced in Section 9.2. Recall that under Brownian motion the expected change along a branch is zero, and the variance of the change grows in proportion to the branch length. The observed trait values at the tips of the tree follow a multivariate normal distribution whose covariance structure is set by the shared branch lengths of the tree.

Under Brownian motion the estimate of the trait value at an internal node has a satisfying intuition. It is a weighted average of the values in the two subtrees descending from the node, where the weights are inversely proportional to the branch lengths (including the variance accumulated deeper in each subtree). A descendant reached by a short branch is a more reliable guide to the ancestor than one reached by a long branch, and so it counts for more in the average. The reconstructed value at the root is a weighted average of the entire tree.

Concretely, the estimate is built up from the tips toward the root by the same kind of pruning we used for the discrete case. Summarize each tip by its observed value and a variance equal to its branch length (in units of the rate σ^2). At an internal node joining two descendants that carry estimates $\hat{x}_i$ and $\hat{x}_j$ with variances v_i and v_j, combine them by inverse-variance weighting,

$$\hat{x} = \frac{\hat{x}_i/v_i + \hat{x}_j/v_j}{1/v_i + 1/v_j}, \tag{11.1}$$

so that the descendant reached by the shorter, lower-variance branch pulls the estimate more strongly toward itself. The combined estimate carries its own variance, $1/(1/v_i + 1/v_j)$, to which we add the node's branch length before passing it further toward the root. Sweeping all the way to the root yields its maximum likelihood estimate together with the variance that becomes its confidence interval. The estimate at any other internal node follows from the same rule applied on both sides of the node—equivalently, from re-rooting the tree at that node and reading off the new root estimate—so it reflects the descendants below and the rest of the tree above in the same inverse-variance way.

Figure 11.2 shows a continuous reconstruction, with the estimated trait value mapped as a grayscale gradient along the branches of the tree. Because the reconstruction at each node is an average of its descendants, the reconstructed values are smoother and less extreme than the tip values. Deep nodes in particular are pulled toward the middle of the observed range. This is a direct consequence of the model. Brownian motion has no memory of direction and no tendency toward any particular value, so in the absence of information the best guess for an ancestor is the average of its descendants.

Just as with discrete characters, we can quantify the uncertainty in these estimates, and the same pruning that produces each estimate also produces its variance. That variance—the $1/(1/v_i + 1/v_j)$ term we tracked up the tree in

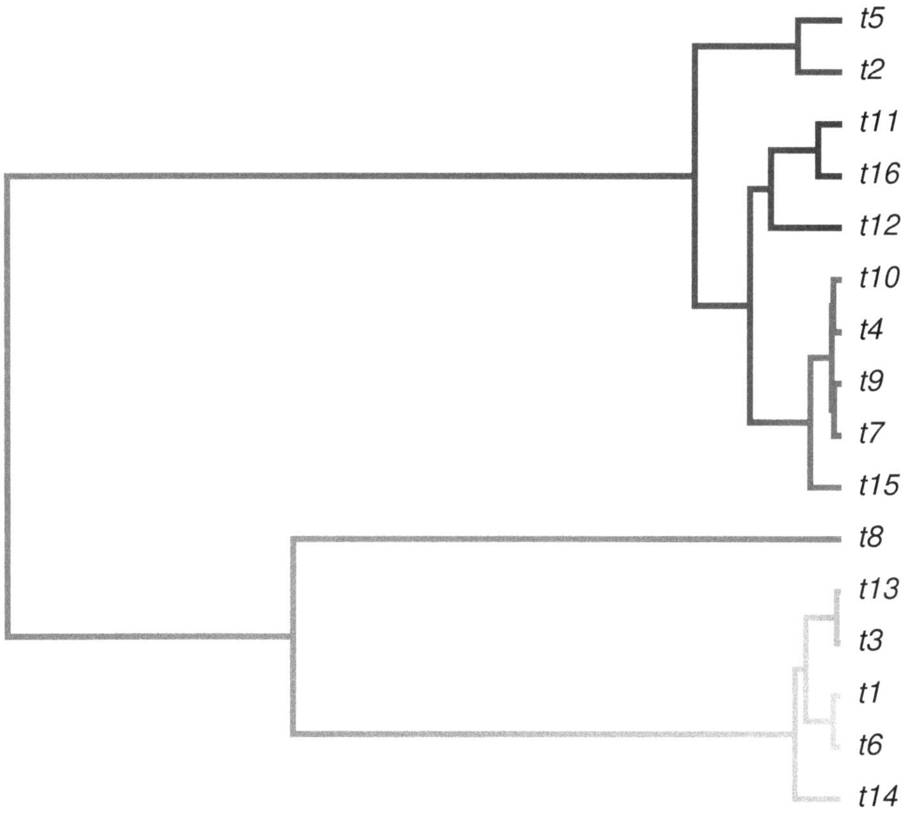

Figure 11.2: Maximum likelihood reconstruction of a continuous character evolving under Brownian motion. A grayscale gradient (light = low, dark = high) maps trait values along the branches, interpolating between the estimated ancestral values at the nodes and the observed values at the tips. The reconstruction is a smooth, branch-length-weighted interpolation, and estimates at deep nodes are pulled toward the overall mean of the tree.

Equation (11.1)—is small when a node is pinned down by short branches to informative tips and large when the node is reachable only across long branches. Multiplying it by the estimated diffusion rate σ^2 gives the variance of the estimate itself, and because trait values are normally distributed under Brownian motion, an approximate 95% confidence interval is the estimate plus or minus about two standard errors (twice the square root of that variance). Interval width therefore grows for nodes that are deeper in the tree and further from the observed tips, whose estimates pool information across longer, higher-variance paths. The reconstruction at the root is often accompanied by a strikingly wide interval, which again is the model being honest—a single trait diffusing randomly for a long time tells us very little about where it started.

11.3 Cautions and limitations

Reconstructions are only as good as the model (Cunningham et al., 1998). Every reconstructed state is conditional on the model of evolution we assumed, the tree we clamped, and the branch lengths we used. Change the model and the reconstruction can change, and reconstructions can be strikingly sensitive to this—a badly misspecified model can yield confident but wrong ancestral states (Revell, 2025). A reconstruction is best understood as a statement of the form "if the character evolved like this, on this tree, then the ancestor probably looked like this."

Uncertainty grows with depth. Both the discrete pies and the continuous confidence intervals tend to become less informative as we move away from the tips and toward the root. Deep ancestral states, which are often the ones we most want to know, are frequently the ones we can say the least about.

Time-reversible models cannot detect directional trends. Many standard models, including Brownian motion and the reversible substitution models, are symmetric with respect to the direction of change. If a trait actually evolved with a consistent directional bias—say, a lineage-wide trend toward larger body size—a reversible model applied only to the surviving tips can be systematically misled about ancestral values (Schluter et al., 1997). Fossil data, when available, can constrain reconstructions in ways that living tips alone cannot.

Chapter 12

Comparative methods

In the previous chapter we used the tree to look back in time and estimate the states of ancestors. Here we use the tree for a different purpose: rather than reconstructing particular historical states, we ask general questions about how traits evolve. This is the domain of phylogenetic comparative methods, one of the largest and fastest-growing areas of phylogenetic biology (Chapter 1). Its questions are wide-ranging, including whether two traits tend to change together, whether the rate of evolution differs across the tree, whether a trait is drawn toward some optimum, and many more topics. This chapter concentrates on the first and most common of them, the evolutionary correlation between two traits.

A recurring theme of phylogenetic comparative biology is that we cannot treat the tips of a tree as independent observations. Species are connected by a shared evolutionary history, and that history leaves a very specific statistical imprint on their trait values. Ignoring it leads to wrong conclusions and limits what questions can even be asked.

12.1 Species are not independent observations

Suppose we measure two traits across a set of species, say body mass and metabolic rate, and we want to know whether they tend to evolve together. The naive approach is to treat each species as an independent data point, plot one trait against the other, and fit an ordinary regression. This is exactly what we would do with independent samples in an introductory statistics course.

The problem is that species are not independent samples. Two species that diverged from a common ancestor a short time ago have had very little time to evolve apart, and so they tend to resemble each other on essentially every trait, simply because they inherited most of their features from that recent common ancestor. Two species whose lineages separated hundreds of millions of years ago have had far more opportunity to diverge. The tree tells us how much shared history each pair of species has, and therefore how much we should expect them to resemble one another before we invoke any relationship between the traits themselves.

This was the central insight of a landmark paper by Joe Felsenstein (Felsenstein, 1985b). If we ignore the tree, we are effectively counting closely related species as if they were independent replicates, when in reality they are close to being a single replicate observed several times. Our sample size is inflated, our confidence intervals are too narrow, and we will routinely find "significant" relationships that are nothing more than the shadow of shared ancestry.

12.1.1 A simulation of the problem

We can see the problem directly by simulating traits that we know are not evolutionarily related, and then asking what an analysis that ignores the tree would conclude. To do this we need two trees that share the same tips but differ in how much history those tips share.

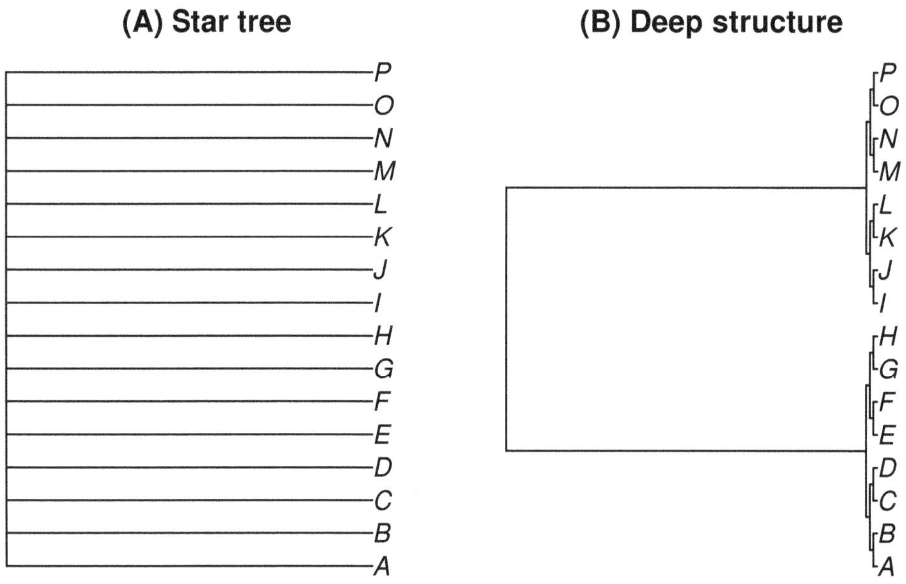

Figure 12.1: Two trees with the same sixteen tips, each scaled to the same root age. (A) A star tree, in which all of the evolutionary change happens on the terminal branches, so the tips share essentially no history. (B) A tree with deep structure, in which most of the change is shared along internal branches, so closely related tips share most of their history.

Figure 12.1 shows two trees over the same sixteen tips. In the star tree, every lineage descends independently from the root, so no two tips share any history beyond the root itself. In the tree with deep structure, most of the branch length is internal and shared, so sister tips are nearly identical by inheritance.

Now we simulate two traits on each tree under Brownian motion (Section 9.2), with no evolutionary covariance between them. Any apparent relationship between the two traits at the tips must therefore be an artifact of the tree, not a real association.

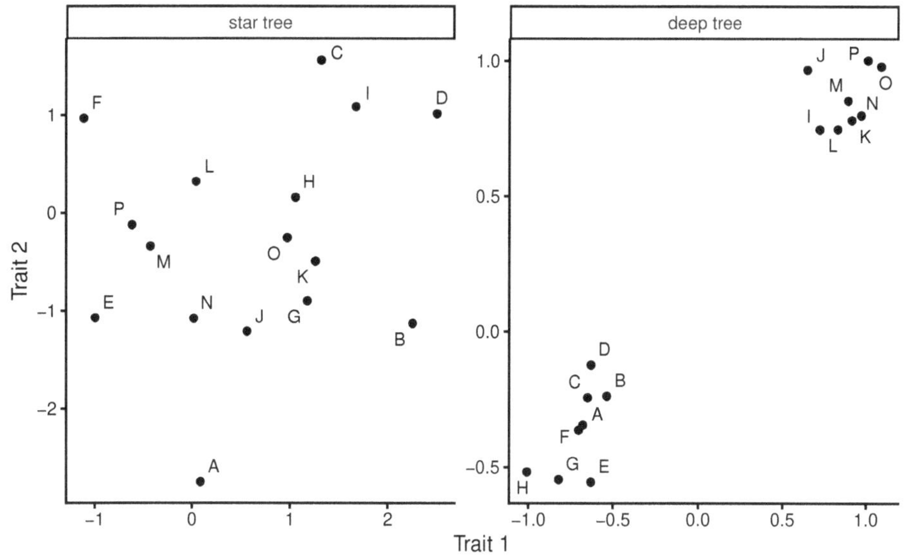

Figure 12.2: Two traits simulated independently under Brownian motion, with no evolutionary covariance between them, shown at the tips of the star tree and the deep tree. On the star tree the tips form an uncorrelated cloud, as they should for independent traits. On the deep tree, closely related tips (adjacent letters) cluster together, and this clustering can produce a strong apparent relationship between the two traits even though none exists.

Figure 12.2 shows the result. On the star tree, where the tips share no history, the two independent traits form the uncorrelated cloud we would expect. On the deep tree, sister tips (adjacent letters, such as A and B, or O and P) cluster tightly together in both traits at once, because they inherited both trait values from a recent common ancestor. These clusters can line up to suggest a strong relationship between the two traits even though we simulated them to be completely independent. An ordinary regression on the deep-tree tips, treating each species as an independent point, would happily report a relationship that does not exist. This is the phenomenon Felsenstein warned about, and it is why we need methods that account for the tree.

12.2 Phylogenetic independent contrasts

Felsenstein's own solution to the problem was elegant, and it remains one of the clearest ways to understand and apply phylogenetic comparative methods (Felsenstein, 1985b). Rather than write the non-independence down and correct for it all at once, he worked directly on the tree, from the tips toward the root, re-expressing the correlated tip values as a set of quantities that are independent under Brownian motion. The method builds on the ancestral character state reconstruction of the previous chapter.

Consider two lineages, i and j, that descend from the same node, with observed trait values x_i and x_j and branch lengths v_i and v_j. Their *contrast* is the standardized difference between them:

$$c = \frac{x_i - x_j}{\sqrt{v_i + v_j}} \tag{12.1}$$

The numerator, $x_i - x_j$, is the key. Because both lineages inherited the same value from their shared ancestor, that ancestral value cancels in the difference, leaving only the independent evolution that happened along the two branches since they diverged. The denominator standardizes the contrast: under Brownian motion the variance of the difference is proportional to $v_i + v_j$, so dividing by $\sqrt{v_i + v_j}$ puts contrasts taken across short branches and long branches on a common scale. The standardized contrasts then all have the same variance, and they are independent of one another.

Having taken the contrast at a node, Felsenstein pruned the two descendants from the tree and replaced them with their ancestor. He estimated the ancestor's trait value as the inverse-branch-length-weighted average of its two descendants, $(x_i/v_i + x_j/v_j)/(1/v_i + 1/v_j)$—the closer descendant counting for more—which is exactly the Brownian-motion ancestral state estimate (Equation (11.1)). Because that ancestor is now an estimate rather than an observation, its own branch is lengthened slightly, by $v_i v_j/(v_i + v_j)$, to carry the extra uncertainty forward. With the pair collapsed to a single reconstructed tip, the same operation applies at the next node down, and so on until the root.

A rooted tree with n tips has $n - 1$ internal nodes and therefore yields $n - 1$ contrasts, one fewer than the number of species. This is the appropriate sample size for an evolutionary question. It reflects the number of independent opportunities for change on the tree, rather than the number of species, which overcounts because of shared ancestry.

To test whether two traits are evolutionarily related, we compute the contrasts for each trait separately and examine the relationship between the two sets. Because the contrasts are independent, an ordinary regression of one set on the other is now valid. By convention it is fit through the origin: the direction in which we happen to subtract each pair of descendants is arbitrary, so a contrast and its negative are equally valid, and only a line through the origin respects that symmetry.

Figure 12.3 shows contrasts for two traits that were simulated with a genuine evolutionary covariance between them. Here the relationship in the contrasts is real, and the regression through the origin recovers it. The same procedure applied to the independent traits of Figure 12.2 would show no such relationship, because the shared ancestry that produced the spurious clustering has been differenced away.

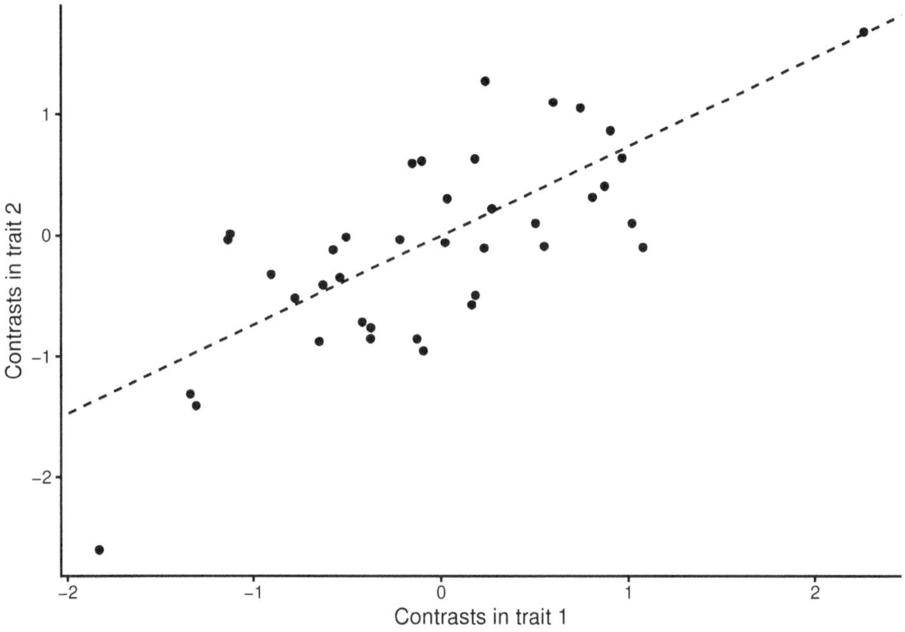

Figure 12.3: Phylogenetic independent contrasts for two traits, computed on a tree with strong structure. Each point is a contrast at one internal node, comparing the two lineages that descend from it. Because the contrasts are independent under Brownian motion, an ordinary regression through the origin (dashed line) is a valid test of whether the two traits are evolutionarily related.

12.3 The expected covariance structure

Independent contrasts handle shared history by transforming the correlated tip values into a set of contrasts that are independent under Brownian motion. There is another, complementary perspective. Instead of transforming the data to independence, it describes the shared history explicitly, as a covariance matrix among the tips.

Under Brownian motion, the trait values at the tips of a tree follow a multivariate normal distribution. The mean of that distribution is the root state, and the covariance between any two tips is proportional to the amount of evolutionary time they share, *i.e.*, the distance from the root to their most recent common ancestor. This is intuitive: two tips accumulate the same random changes along every branch they share, and independent changes only after they diverge, so the more history they share, the more their trait values covary. A tip shares its entire root-to-tip path with itself, so its variance (its covariance with itself) is proportional to its total distance from the root.

We can read this expected covariance structure directly off the tree. Consider the small four-tip tree in Figure 12.4, drawn with its branch lengths.

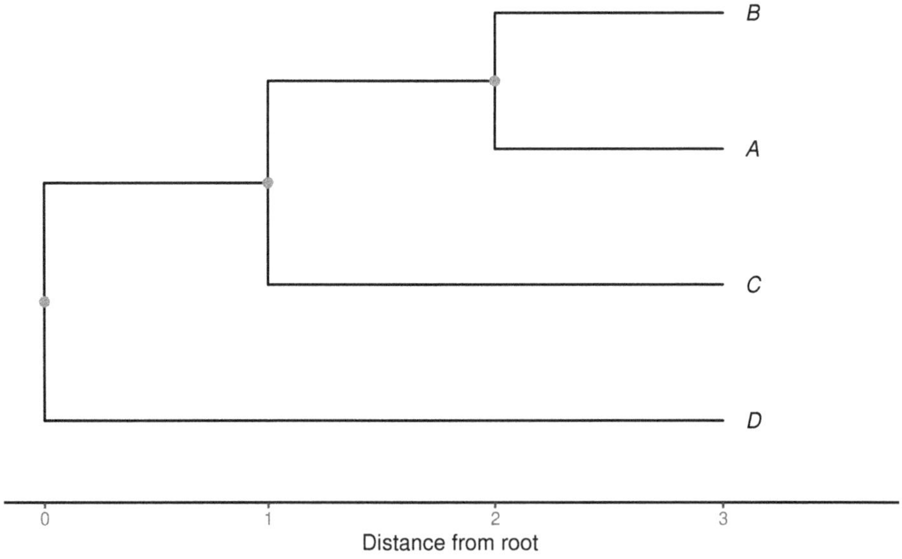

Figure 12.4: A four-tip tree drawn with its branch lengths; the horizontal axis is distance from the root. The covariance between two tips is the distance from the root to their most recent common ancestor—the internal node (gray) at which their lineages join. Sisters A and B join at the more recent node (distance 2), A and C (and B and C) join at the older node (distance 1), and D joins the others only at the root (distance 0). The variance of each tip is its total distance from the root, here 3 for every tip because the tree is ultrametric. These depths are exactly the entries of the covariance matrix in Equation (12.2).

To turn this tree into a matrix, we take the tips two at a time and measure how much evolutionary history each pair shares. There are two kinds of entry. A *diagonal* entry is the variance of a single tip, and equals its total distance from the root—the full length of the path from the root out to that tip. An *off-diagonal* entry is the covariance between two different tips, and equals the length of the path they share: the distance from the root to their most recent common ancestor, the node at which their two lineages join.

Working through the tips of Figure 12.4:

- Every tip sits at distance 3 from the root, so every diagonal entry, the variance, is 3.
- A and B are sisters. Their lineages join at the more recent internal node, at distance 2 from the root, so the path they share—and therefore their covariance—is 2.
- A and C join further back, at the older internal node at distance 1 from the root, so their covariance is 1. The same holds for B and C.
- D branches off at the root itself, so it shares no path with A, B, or C, and all of its covariances are 0.

Collecting these values gives the phylogenetic variance-covariance matrix. We will call it $\mathbf{\Sigma}$, a symbol kept deliberately distinct from the tip names, with one row and one column for each tip:

$$\mathbf{\Sigma} = \begin{array}{c} \\ A \\ B \\ C \\ D \end{array} \begin{array}{c} \begin{array}{cccc} A & B & C & D \end{array} \\ \begin{pmatrix} 3 & 2 & 1 & 0 \\ 2 & 3 & 1 & 0 \\ 1 & 1 & 3 & 0 \\ 0 & 0 & 0 & 3 \end{pmatrix} \end{array} \tag{12.2}$$

The diagonal, running from top left to bottom right, holds the variances, all equal here because the tree is ultrametric. Every other entry is a covariance, and the pattern of large and small values mirrors the tree: closely related tips such as A and B share a great deal of history and have a large covariance, while the distantly attached D shares none and has covariances of zero. Reading across the row for D, or down its column, gives all zeros. This matrix is a complete description of the non-independence induced by the tree, and it is exactly what the next method takes as its input.

12.4 Phylogenetic generalized least squares

Independent contrasts turn out to be one instance of a more general and flexible framework. That framework is phylogenetic generalized least squares, abbreviated PGLS (Grafen, 1989; Freckleton et al., 2002; Symonds and Blomberg, 2014).

Ordinary least squares regression assumes that the residuals, the deviations of the data from the fitted line, are independent and identically distributed. Generalized least squares relaxes this assumption by allowing the residuals to have a specified covariance structure. And we have just built exactly such a structure: the phylogenetic variance-covariance matrix $\mathbf{\Sigma}$ of Equation (12.2). In PGLS we fit a regression in which the residuals are assumed to covary according to $\mathbf{\Sigma}$. The method effectively down-weights the contribution of species that share a great deal of history, so that a cluster of closely related species counts for less than the same number of independent lineages.

```
pgls_slope contrasts_slope
 0.7384165       0.7384165
```

Independent contrasts are the special case of PGLS in which the covariance structure $\mathbf{\Sigma}$ is the one implied by Brownian motion. For that model the two methods are mathematically equivalent: the slope from the PGLS regression is identical to the slope from regressing the contrasts through the origin. Felsenstein's node-by-node pruning and the matrix-based regression are two routes to the same estimate.

Because PGLS is more general, it has become the more widely used of the two. Expressed as a regression with a specified covariance structure, it accommodates things that are awkward under the contrasts formulation. We can include multiple predictors, categorical variables, and continuous covariates in the same model. More importantly, we can change the assumed covariance structure to reflect a different model of evolution. If we think the trait evolves under an Ornstein-Uhlenbeck process (Section 9.2) (Hansen, 1997; Butler and King, 2004) rather than pure Brownian motion, we substitute the covariance structure implied by that model. The tree still supplies the non-independence, but we now have a whole family of models describing how that non-independence accumulates.

It is worth being precise about what these methods do, because they are easily misdescribed (Uyeda et al., 2018). It is common to say that they "remove the phylogenetic signal" (Blomberg et al., 2003) or "correct for the fact that species are not independent," but neither phrasing is quite right. Ordinary regression does not assume that the observations themselves are independent; it assumes that the *residuals*—the part of one trait left unexplained by the other—are independent. What independent contrasts and PGLS do is account for the covariance of those residuals, and they discard no biological signal in doing so; the contrasts are themselves the evolutionary changes. A subtler and more important caution is that correctly accounting for shared history does not, by itself, show that two traits are causally related. A single unreplicated event—one ancient branch on which both traits happened to shift—can produce a strongly significant contrast or PGLS slope even when nothing connects the traits, and neither method guards against this. Shared history is not just a statistical nuisance to be swept away.

12.5 Beyond correlated evolution

Testing for correlated evolution between two traits is the most common comparative analysis, and it is where the core ideas are clearest, but it is far from the only question comparative methods can address (Harmon, 2018; Revell and

Harmon, 2022). The same framework of an explicit evolutionary model on a fixed tree supports a much broader set of investigations.

We can ask whether the rate of evolution itself varies across the tree, for example whether a particular clade diversified in a trait much faster than its relatives. We can ask whether a trait is subject to stabilizing selection toward an optimum, using the Ornstein-Uhlenbeck model, and whether different parts of the tree are drawn toward different optima. We can fit and compare these competing models using the model evaluation tools developed earlier in the book, asking which model of trait evolution the data support. In every case the logic is the one we have built up here: specify a model of how the trait evolves, use the tree to describe the resulting non-independence among species, and then estimate parameters or compare models in that light.

Appendix A

Statistical and mathematical fundamentals

Phylogenetic biology draws on a variety of themes in statistics and mathematics. Given the diverse backgrounds of people coming to phylogenetic biology, one of the challenges of introducing people to this field is ensuring a foundation in these fundamentals. Many people have a solid working grasp of these topics, others need a refresher, and some have never seen them.

Throughout the text I have covered some of these fundamentals, but here I provide additional resources.

A.1 Probability theory and general statistics

I put together this short summary of statistical concepts commonly encountered in biology—https://raw.githubusercontent.com/caseywdunn/statistics/master/statistics_in_biology.pdf. It is short and technical, with a focus on the relationship between concepts, such as the exponential family of distributions.

A.2 Bayes' theorem

Grant Sanderson of 3Blue1Brown provides an excellent introduction to Bayes' theorem and conditional probability in this video—https://youtu.be/HZGCoVF3YvM. Note that he uses E for Evidence rather than D for Data as is common practice in phylogenetic applications.

This video provides additional perspective, including its application to medical testing—https://youtu.be/R13BD8qKeTg.

The MCMC robot by Paul Lewis is an excellent interactive introduction to MCMC sampling—https://phylogeny.uconn.edu/mcmc-robot/.

A.3 Linear algebra

Linear algebra is a rich field of math that springs from a surprisingly simple starting point—linear combinations of vectors, often represented as rows and columns of matrices. From basic operations such as vector addition and scalar multiplication, one can build matrix operations and many sophisticated data analyses. Many familiar data analysis methods, such as regression, solving systems of equations, Principal Component Analysis (PCA), and common clustering methods, are built from linear algebra under the hood. So are many aspects of phylogenetic analysis, including the models of evolution themselves and the representation of the phylogenetic trees in matrix forms that computers can easily manipulate.

Though linear algebra is fundamental to many topics at the intersection of science, mathematics, and computation, most science students have far less exposure to this critical field than to statistics, calculus, and other quantitative topics. This is too bad, as even a basic working knowledge of linear algebra can provide great insight into the tools scientists use daily and allow us to better apply, implement, and extend them.

The following resources are a good place to start in your own studies of linear algebra:

- The *Essence of linear algebra* video series by Grant Sanderson (3Blue1Brown)—https://www.3blue1brown.com/topics/linear-algebra.
- *Introduction to Linear Algebra* by Gilbert Strang (Strang, 2023) is a widely used and comprehensive textbook, paired with his well-known MIT lecture series.
- Linear Algebra: Theory, Intuition, Code, by Mike X Cohen. This book is accessible to scientists with little past math experience, and is very reasonably priced. https://www.amazon.com/Linear-Algebra-Theory-Intuition-Code/dp/9083136604.
- Data Driven Science and Engineering, by Steve Brunton and Nathan Kutz—http://www.databookuw.com/. This book shows how linear algebra is applied to many powerful scientific data analyses. The authors have excellent companion videos organized into playlists at https://www.youtube.com/c/Eigensteve/playlists, including singular value decomposition for Chapter 1, and Fourier analyses for Chapter 2.

Appendix B

Software versions

This book, version 1.0.0, was rendered from the source code on Jul 26, 2026 at 03:28:37 AM, at git commit 7702033 (2026-07-25), with the following R package versions.

```
R version 4.5.1 (2025-06-13)
Platform: x86_64-pc-linux-gnu
Running under: Ubuntu 24.04.3 LTS

Matrix products: default
BLAS:   /usr/lib/x86_64-linux-gnu/openblas-pthread/libblas.s
o.3
LAPACK: /usr/lib/x86_64-linux-gnu/openblas-pthread/libopenbl
asp-r0.3.26.so;  LAPACK version 3.12.0

locale:
 [1] LC_CTYPE=en_US.UTF-8       LC_NUMERIC=C
 [3] LC_TIME=en_US.UTF-8        LC_COLLATE=en_US.UTF-8
 [5] LC_MONETARY=en_US.UTF-8    LC_MESSAGES=en_US.UTF-8
 [7] LC_PAPER=en_US.UTF-8       LC_NAME=C
 [9] LC_ADDRESS=C               LC_TELEPHONE=C
[11] LC_MEASUREMENT=en_US.UTF-8 LC_IDENTIFICATION=C

time zone: Etc/UTC
tzcode source: system (glibc)

attached base packages:
[1] stats     graphics  grDevices utils     datasets
[6] methods   base

other attached packages:
 [1] nlme_3.1-168     scales_1.4.0     ggrepel_0.9.6
 [4] kableExtra_1.4.0 phangorn_2.12.1  Matrix_1.7-3
 [7] lubridate_1.9.4  forcats_1.0.1    dplyr_1.1.4
```

```
[10] purrr_1.1.0      readr_2.1.5      tidyr_1.3.1
[13] tibble_3.3.0     ggplot2_4.0.0    tidyverse_2.0.0
[16] stringr_1.5.2    magrittr_2.0.4   gridExtra_2.3
[19] geiger_2.0.11    phytools_2.5-2   maps_3.4.3
[22] ape_5.8-1        ggtree_4.0.5     treeio_1.34.0
[25] renv_1.1.8       bookdown_0.45

loaded via a namespace (and not attached):
 [1] mnormt_2.1.1            rlang_1.1.6
 [3] compiler_4.5.1          mgcv_1.9-3
 [5] png_0.1-8               systemfonts_1.3.1
 [7] vctrs_0.6.5             combinat_0.0-8
 [9] quadprog_1.5-8          crayon_1.5.3
[11] pkgconfig_2.0.3         fastmap_1.2.0
[13] magick_2.9.1            labeling_0.4.3
[15] subplex_1.9             deSolve_1.40
[17] rmarkdown_2.30          tzdb_0.5.0
[19] bit_4.6.0               tinytex_0.57
[21] xfun_0.53               cachem_1.1.0
[23] aplot_0.2.9             clusterGeneration_1.3.8
[25] jsonlite_2.0.0          parallel_4.5.1
[27] R6_2.6.1                bslib_0.9.0
[29] stringi_1.8.7           RColorBrewer_1.1-3
[31] jquerylib_0.1.4         numDeriv_2016.8-1.1
[33] Rcpp_1.1.0              iterators_1.0.14
[35] knitr_1.50              optimParallel_1.0-2
[37] splines_4.5.1           igraph_2.2.1
[39] timechange_0.3.0        tidyselect_1.2.1
[41] rstudioapi_0.17.1       yaml_2.3.10
[43] doParallel_1.0.17       codetools_0.2-20
[45] lattice_0.22-7          withr_3.0.2
[47] S7_0.2.0                coda_0.19-4.1
[49] evaluate_1.0.5          ggimage_0.3.5
[51] gridGraphics_0.5-1      xml2_1.4.1
[53] pillar_1.11.1           foreach_1.5.2
[55] ggfun_0.2.0             generics_0.1.4
[57] vroom_1.6.6             hms_1.1.4
[59] tidytree_0.4.6          glue_1.8.0
[61] gdtools_0.4.4           scatterplot3d_0.3-44
[63] lazyeval_0.2.2          tools_4.5.1
[65] ggiraph_0.9.2           fs_1.6.6
[67] mvtnorm_1.3-3           fastmatch_1.1-6
[69] grid_4.5.1              patchwork_1.3.2
[71] cli_3.6.5               rappdirs_0.3.3
[73] DEoptim_2.2-8           textshaping_1.0.4
[75] fontBitstreamVera_0.1.1 expm_1.0-0
[77] viridisLite_0.4.2       svglite_2.2.2
[79] gtable_0.3.6            yulab.utils_0.2.1
```

```
[81] sass_0.4.10             digest_0.6.37
[83] fontquiver_0.2.1        ggplotify_0.1.3
[85] htmlwidgets_1.6.4       farver_2.1.2
[87] htmltools_0.5.8.1       lifecycle_1.0.4
[89] fontLiberation_0.1.0    bit64_4.6.0-1
[91] MASS_7.3-65
```

Bibliography

Hirotugu Akaike. A new look at the statistical model identification. *IEEE Transactions on Automatic Control*, 19(6):716–723, 1974. doi:10.1109/TAC.1974.1100705.

David A. Baum and Stacey D. Smith. *Tree Thinking: An Introduction to Phylogenetic Biology*. Roberts and Company Publishers, Greenwood Village, CO, 2012.

Simon P. Blomberg, Theodore Garland, and Anthony R. Ives. Testing for phylogenetic signal in comparative data: Behavioral traits are more labile. *Evolution*, 57(4):717–745, 2003. doi:10.1111/j.0014-3820.2003.tb00285.x.

George EP Box. Science and statistics. *Journal of the American Statistical Association*, 71(356):791–799, 1976. doi:10.1080/01621459.1976.10480949.

Leo Breiman. Statistical modeling: The two cultures (with comments and a rejoinder by the author). *Statistical Science*, 16(3):199–231, 2001. doi:10.1214/ss/1009213726. URL https://projecteuclid.org/euclid.ss/1009213726.

KP Burnham and DR Anderson. *Model Selection and Multimodel Inference: A Practical Information-Theoretic Approach*. Springer, 2nd edition, 2002. doi:10.1007/b97636.

Marguerite A. Butler and Aaron A. King. Phylogenetic comparative analysis: A modeling approach for adaptive evolution. *The American Naturalist*, 164 (6):683–695, 2004. doi:10.1086/426002.

E Sally Chang, Moran Neuhof, Nimrod D Rubinstein, Arik Diamant, Hervé Philippe, Dorothée Huchon, and Paulyn Cartwright. Genomic insights into the evolutionary origin of Myxozoa within Cnidaria. *Proceedings of the National Academy of Sciences*, 112(48):14912–14917, 2015. doi:10.1073/pnas.1511468112.

Kevin Chen, Dannie Durand, and Martin Farach-Colton. Notung: dating gene duplications using gene family trees. In *Proceedings of the Fourth Annual International Conference on Computational Molecular Biology (RECOMB '00)*, RECOMB '00, pages 96–106, New York, NY, USA, 2000. Association for Computing Machinery. ISBN 1581131860. doi:10.1145/332306.332351.

Ming-Hui Chen, Lynn Kuo, and Paul O Lewis, editors. *Bayesian Phylogenetics: Methods, Algorithms, and Applications.* CRC Press, 2014. doi:10.1201/b16965.

Julia Chifman and Laura Kubatko. Quartet inference from SNP data under the coalescent model. *Bioinformatics*, 30(23):3317–3324, 2014. doi:10.1093/bioinformatics/btu530.

Clifford W. Cunningham, Kevin E. Omland, and Todd H. Oakley. Reconstructing ancestral character states: a critical reappraisal. *Trends in Ecology & Evolution*, 13(9):361–366, 1998. doi:10.1016/S0169-5347(98)01382-2.

Lucas Czech, Jaime Huerta-Cepas, and Alexandros Stamatakis. A Critical Review on the Use of Support Values in Tree Viewers and Bioinformatics Toolkits. *Molecular Biology and Evolution*, 34(6):1535–1542, 03 2017. ISSN 0737-4038. doi:10.1093/molbev/msx055. URL https://doi.org/10.1093/molbev/msx055.

Charles Darwin. *The origin of species by means of natural selection: or, the preservation of favored races in the struggle for life.* John Murray, London, 1859.

Kevin de Queiroz. Toward an integrated system of clade names. *Systematic Biology*, 56(6):956–974, 2007. doi:10.1080/10635150701656378.

Kevin de Queiroz and Philip D Cantino. *International Code of Phylogenetic Nomenclature (PhyloCode): A Phylogenetic Code of Biological Nomenclature.* CRC Press, 2020. doi:10.1201/9780429446320.

Alexei J. Drummond and Andrew Rambaut. BEAST: Bayesian evolutionary analysis by sampling trees. *BMC Evolutionary Biology*, 7:214, 2007. doi:10.1186/1471-2148-7-214.

Alexei J. Drummond, Simon Y. W. Ho, Matthew J. Phillips, and Andrew Rambaut. Relaxed phylogenetics and dating with confidence. *PLoS Biology*, 4(5):e88, 2006. doi:10.1371/journal.pbio.0040088.

Casey W. Dunn and Catriona Munro. Comparative genomics and the diversity of life. *Zoologica Scripta*, 45(S1):5–13, 2016. doi:10.1111/zsc.12211.

Casey W Dunn, Felipe Zapata, Catriona Munro, Stefan Siebert, and Andreas Hejnol. Pairwise comparisons across species are problematic when analyzing functional genomic data. *Proceedings of the National Academy of Sciences*, 115(3):E409–E417, 2018. doi:10.1073/pnas.1707515115.

Robert C. Edgar. MUSCLE: multiple sequence alignment with high accuracy and high throughput. *Nucleic Acids Research*, 32(5):1792–1797, 2004. doi:10.1093/nar/gkh340.

Scott V. Edwards. Is a new and general theory of molecular systematics emerging? *Evolution*, 63(1):1–19, 2009. doi:10.1111/j.1558-5646.2008.00549.x.

David M. Emms and Steven Kelly. OrthoFinder: phylogenetic orthology inference for comparative genomics. *Genome Biology*, 20(1):238, 2019. doi:10.1186/s13059-019-1832-y.

J. Felsenstein. Evolutionary trees from DNA sequences: a maximum likelihood approach. *Journal of Molecular Evolution*, 17(6):368–376, 1981. doi:10.1007/BF01734359. URL http://eutils.ncbi.nlm.nih.gov/entrez/eutils/elink.fcgi?dbfrom=pubmed&id=7288891&retmode=ref&cmd=prlinks.

Joseph Felsenstein. Maximum-likelihood estimation of evolutionary trees from continuous characters. *American Journal of Human Genetics*, 25(5): 471–492, 1973.

Joseph Felsenstein. Confidence limits on phylogenies: an approach using the bootstrap. *Evolution*, 39(4):783–791, 1985a. doi:10.1111/j.1558-5646.1985.tb00420.x.

Joseph Felsenstein. Phylogenies and the comparative method. *The American Naturalist*, 125(1):1–15, 1985b. doi:10.1086/284325.

Joseph Felsenstein. *Inferring Phylogenies.* Sinauer Associates, Sunderland, MA, 2004.

Walter M. Fitch. Toward defining the course of evolution: Minimum change for a specific tree topology. *Systematic Zoology*, 20(4):406–416, 1971. doi:10.1093/sysbio/20.4.406.

Robert P. Freckleton, Paul H. Harvey, and Mark Pagel. Phylogenetic analysis and comparative data: A test and review of evidence. *The American Naturalist*, 160(6):712–726, 2002. doi:10.1086/343873.

Peter Godfrey-Smith. *Philosophy of biology.* Princeton University Press, 2014. doi:10.1515/9781400850440.

Nick Goldman, Jon P. Anderson, and Allen G. Rodrigo. Likelihood-based tests of topologies in phylogenetics. *Systematic Biology*, 49(4):652–670, 12 2000. ISSN 1063-5157. doi:10.1080/106351500750049752. URL https://doi.org/10.1080/106351500750049752.

Alan Grafen. The phylogenetic regression. *Philosophical Transactions of the Royal Society of London. B, Biological Sciences*, 326(1233):119–157, 1989. doi:10.1098/rstb.1989.0106.

James Hadfield, Colin Megill, Sidney M Bell, John Huddleston, Barney Potter, Charlton Callender, Pavel Sagulenko, Trevor Bedford, and Richard A Neher. Nextstrain: real-time tracking of pathogen evolution. *Bioinformatics*, 34 (23):4121–4123, 2018. doi:10.1093/bioinformatics/bty407.

Ernst Haeckel. *Generelle Morphologie der Organismen.* Georg Reimer, Berlin, 1866.

Ernst Haeckel. *The evolution of man*, volume 1. D. Appleton, New York, 1897.

Thomas F. Hansen. Stabilizing selection and the comparative analysis of adaptation. *Evolution*, 51(5):1341–1351, 1997. doi:10.1111/j.1558-5646.1997.tb01457.x.

Luke Harmon. Phylogenetic comparative methods: learning from trees. 2018. doi:10.32942/osf.io/e3xnr. URL https://lukejharmon.github.io/pcm/.

Masami Hasegawa, Hirohisa Kishino, and Taka-aki Yano. Dating of the human-ape splitting by a molecular clock of mitochondrial DNA. *Journal of Molecular Evolution*, 22(2):160–174, 1985. doi:10.1007/BF02101694.

W. K. Hastings. Monte Carlo sampling methods using Markov chains and their applications. *Biometrika*, 57(1):97–109, 1970. doi:10.1093/biomet/57.1.97.

Tracy A Heath, John P Huelsenbeck, and Tanja Stadler. The fossilized birth–death process for coherent calibration of divergence-time estimates. *PNAS*, 111(29):E2957–E2966, 2014. doi:10.1073/pnas.1319091111.

Joseph Heled and Alexei J. Drummond. Bayesian inference of species trees from multilocus data. *Molecular Biology and Evolution*, 27(3):570–580, 2010. doi:10.1093/molbev/msp274.

Willi Hennig. *Phylogenetic Systematics*. University of Illinois Press, 1966.

Diep Thi Hoang, Olga Chernomor, Arndt von Haeseler, Bui Quang Minh, and Le Sy Vinh. UFBoot2: Improving the ultrafast bootstrap approximation. *Molecular Biology and Evolution*, 35(2):518–522, 2018. doi:10.1093/molbev/msx281.

S. Höhna, T. A. Heath, B. Boussau, M. J. Landis, F. Ronquist, and J. P. Huelsenbeck. Probabilistic graphical model representation in phylogenetics. *Systematic Biology*, 63(5):753–771, 2014. doi:10.1093/sysbio/syu039.

Mark Holder and Paul O Lewis. Phylogeny estimation: traditional and Bayesian approaches. *Nature Reviews Genetics*, 4(4):275–284, 2003. doi:10.1038/nrg1044.

David Houle, Christophe Pélabon, Günter P Wagner, and Thomas F Hansen. Measurement and meaning in biology. *The Quarterly Review of Biology*, 86 (1):3–34, 2011. doi:10.1086/658408.

John P Huelsenbeck and Fredrik Ronquist. MRBAYES: Bayesian inference of phylogenetic trees. *Bioinformatics*, 17(8):754–755, 2001. doi:10.1093/bioinformatics/17.8.754.

Jeffrey B Joy, Richard H Liang, Rosemary M McCloskey, T Nguyen, and Art FY Poon. Ancestral reconstruction. *PLoS Computational Biology*, 12 (7):e1004763, 2016. doi:10.1371/journal.pcbi.1004763.

TH Jukes and CR Cantor. *Evolution of Protein Molecules*. Academic Press New York, NY, 1969. doi:10.1016/B978-1-4832-3211-9.50009-7.

Subha Kalyaanamoorthy, Bui Quang Minh, Thomas K. F. Wong, Arndt von Haeseler, and Lars S. Jermiin. ModelFinder: fast model selection for accurate phylogenetic estimates. *Nature Methods*, 14(6):587–589, 2017. doi:10.1038/nmeth.4285.

Kazutaka Katoh and Daron M. Standley. MAFFT multiple sequence alignment software version 7: Improvements in performance and usability. *Molecular Biology and Evolution*, 30(4):772–780, 2013. doi:10.1093/molbev/mst010.

Motoo Kimura. A simple method for estimating evolutionary rates of base substitutions through comparative studies of nucleotide sequences. *Journal of Molecular Evolution*, 16(2):111–120, 1980. doi:10.1007/BF01731581.

Hirohisa Kishino and Masami Hasegawa. Evaluation of the maximum likelihood estimate of the evolutionary tree topologies from DNA sequence data, and the branching order in Hominoidea. *Journal of Molecular Evolution*, 29(2):170–179, 1989. doi:10.1007/BF02100115.

Michael J Landis, Deren AR Eaton, Wendy L Clement, Brian Park, Elizabeth L Spriggs, Patrick W Sweeney, Erika J Edwards, and Michael J Donoghue. Joint phylogenetic estimation of geographic movements and biome shifts during the global diversification of Viburnum. *Systematic Biology*, 70(1):67–85, 2021. doi:10.1093/sysbio/syaa027.

Bret Larget and Donald L Simon. Markov chain Monte Carlo algorithms for the Bayesian analysis of phylogenetic trees. *Molecular Biology and Evolution*, 16(6):750–759, 1999. doi:10.1093/oxfordjournals.molbev.a026160.

Emily Moriarty Lemmon and Alan R. Lemmon. High-throughput genomic data in systematics and phylogenetics. *Annual Review of Ecology, Evolution, and Systematics*, 44:99–121, 2013. doi:10.1146/annurev-ecolsys-110512-135822.

Paul O. Lewis. A likelihood approach to estimating phylogeny from discrete morphological character data. *Systematic Biology*, 50(6):913–925, 2001. doi:10.1080/106351501753462876.

David R. Maddison, David L. Swofford, and Wayne P. Maddison. NEXUS: An extensible file format for systematic information. *Systematic Biology*, 46(4): 590–621, 1997. doi:10.1093/sysbio/46.4.590.

Wayne P Maddison. Gene trees in species trees. *Systematic Biology*, 46(3): 523–536, 1997. doi:10.1093/sysbio/46.3.523.

Etai Markowski and Edward Susko. Performance of topology tests under extreme selection bias. *Molecular Biology and Evolution*, 41(1):msad280, 1 2024. ISSN 1537-1719. doi:10.1093/molbev/msad280. URL https://doi.org/10.1093/molbev/msad280.

Nicholas Metropolis, Arianna W Rosenbluth, Marshall N Rosenbluth, Augusta H Teller, and Edward Teller. Equation of state calculations by fast computing machines. *The Journal of Chemical Physics*, 21(6):1087–1092, 1953. doi:10.1063/1.1699114.

Bui Quang Minh, Heiko A Schmidt, Olga Chernomor, Dominik Schrempf, Michael D Woodhams, Arndt von Haeseler, and Robert Lanfear. IQ-TREE 2: New models and efficient methods for phylogenetic inference in the genomic era. *Molecular Biology and Evolution*, 37(5):1530–1534, 02 2020. ISSN 0737-4038. doi:10.1093/molbev/msaa015. URL https://doi.org/10.1093/molbev/msaa015.

Lam-Tung Nguyen, Heiko A. Schmidt, Arndt von Haeseler, and Bui Quang Minh. IQ-TREE: A fast and effective stochastic algorithm for estimating maximum-likelihood phylogenies. *Molecular Biology and Evolution*, 32(1): 268–274, 2015. doi:10.1093/molbev/msu300.

Mark Pagel. Inferring the historical patterns of biological evolution. *Nature*, 401(6756):877–884, 1999. doi:10.1038/44766.

Emmanuel Paradis, Simon Blomberg, Ben Bolker, Joseph Brown, Santiago Claramunt, Julien Claude, Hoa Sien Cuong, Richard Desper, Gilles Didier, Benoit Durand, Julien Dutheil, RJ Ewing, Olivier Gascuel, Thomas Guillerme, Christoph Heibl, Anthony Ives, Bradley Jones, Franz Krah, Daniel Lawson, Vincent Lefort, Pierre Legendre, Jim Lemon, Guillaume Louvel, Federico Marotta, Eric Marcon, Rosemary McCloskey, Johan Nylander, Rainer Opgen-Rhein, Andrei-Alin Popescu, Manuela Royer-Carenzi, Klaus Schliep, Korbinian Strimmer, and Damien de Vienne. *ape: Analyses of Phylogenetics and Evolution*, 2024. URL https://github.com/emmanuelparadis/ape. R package version 5.8-1.

James F. Parham, Philip C. J. Donoghue, Christopher J. Bell, Tyler D. Calway, Jason J. Head, Patricia A. Holroyd, Jun G. Inoue, Randall B. Irmis, Walter G. Joyce, Daniel T. Ksepka, José S. L. Patané, Nathan D. Smith, James E. Tarver, Marcel van Tuinen, Ziheng Yang, Kenneth D. Angielczyk, Jenny M. Greenwood, Christy A. Hipsley, Louis Jacobs, Peter J. Makovicky, Johannes Müller, Krister T. Smith, Jessica M. Theodor, Rachel C. M. Warnock, and Michael J. Benton. Best practices for justifying fossil calibrations. *Systematic Biology*, 61(2):346–359, 2012. doi:10.1093/sysbio/syr107.

David Posada and Thomas R Buckley. Model selection and model averaging in phylogenetics: advantages of Akaike information criterion and Bayesian approaches over likelihood ratio tests. *Systematic Biology*, 53(5):793–808, 2004. doi:10.1080/10635150490522304.

David Posada and Keith A. Crandall. Selecting the best-fit model of nucleotide substitution. *Systematic Biology*, 50(4):580–601, 2001. doi:10.1080/106351501750435121.

Liam J. Revell. Ancestral state reconstruction of phenotypic characters. *Evolutionary Biology*, 52(1):1–25, 2025. doi:10.1007/s11692-025-09645-y.

Liam J. Revell and Luke J. Harmon. *Phylogenetic Comparative Methods in R*. Princeton University Press, 2022.

Fredrik Ronquist, Maxim Teslenko, Paul van der Mark, Daniel L. Ayres, Aaron Darling, Sebastian Höhna, Bret Larget, Liang Liu, Marc A. Suchard, and John P. Huelsenbeck. MrBayes 3.2: Efficient Bayesian phylogenetic inference and model choice across a large model space. *Systematic Biology*, 61(3):539–542, 2012. doi:10.1093/sysbio/sys029.

Michael S. Rosenberg, Sankar Subramanian, and Sudhir Kumar. Patterns of Transitional Mutation Biases Within and Among Mammalian Genomes. *Molecular Biology and Evolution*, 20(6):988–993, 06 2003. ISSN 0737-4038. doi:10.1093/molbev/msg113. URL https://doi.org/10.1093/molbev/msg113.

Michael J. Sanderson. A nonparametric approach to estimating divergence times in the absence of rate constancy. *Molecular Biology and Evolution*, 14 (12):1218–1231, 1997. doi:10.1093/oxfordjournals.molbev.a025731.

Dolph Schluter, Trevor Price, Arne Ø. Mooers, and Donald Ludwig. Likelihood of ancestor states in adaptive radiation. *Evolution*, 51(6): 1699–1711, 1997. doi:10.1111/j.1558-5646.1997.tb05095.x.

Gideon Schwarz. Estimating the dimension of a model. *The Annals of Statistics*, 6(2):461–464, 1978. doi:10.1214/aos/1176344136.

Charles Semple and Mike Steel. *Phylogenetics*, volume 24 of *Oxford Lecture Series in Mathematics and Its Applications*. Oxford University Press, 2003. doi:10.1093/oso/9780198509424.001.0001.

H Shimodaira and M Hasegawa. Multiple comparisons of log-likelihoods with applications to phylogenetic inference. *Molecular Biology and Evolution*, 16 (8):1114–1116, 08 1999. ISSN 0737-4038. doi:10.1093/oxfordjournals.molbev.a026201.

Hidetoshi Shimodaira. An approximately unbiased test of phylogenetic tree selection. *Systematic Biology*, 51(3):492–508, 05 2002. ISSN 1063-5157. doi:10.1080/10635150290069913. URL https://doi.org/10.1080/10635150290069913.

Claudia Solís-Lemus, Paul Bastide, and Cécile Ané. PhyloNetworks: a package for phylogenetic networks. *Molecular Biology and Evolution*, 34(12): 3292–3298, 2017. doi:10.1093/molbev/msx235.

Alexandros Stamatakis. RAxML version 8: a tool for phylogenetic analysis and post-analysis of large phylogenies. *Bioinformatics*, 30(9):1312–1313, 2014. doi:10.1093/bioinformatics/btu033.

Mike Steel and David Penny. Parsimony, likelihood, and the role of models in molecular phylogenetics. *Molecular Biology and Evolution*, 17(6):839–850, 2000. doi:10.1093/oxfordjournals.molbev.a026364.

Gilbert Strang. *Introduction to Linear Algebra*. Wellesley-Cambridge Press, 6th edition, 2023.

DL Swofford, GJ Olsen, PJ Waddell, and DM Hillis. *Phylogenetic Inference*. Sinauer Associates Sunderland, MA, 2nd edition, 1996.

Matthew RE Symonds and Simon P Blomberg. A primer on phylogenetic generalised least squares. In László Zsolt Garamszegi, editor, *Modern phylogenetic comparative methods and their application in evolutionary biology*, pages 105–130. Springer, 2014. doi:10.1007/978-3-662-43550-2_5.

Gergely J. Szöllősi, Eric Tannier, Vincent Daubin, and Bastien Boussau. The inference of gene trees with species trees. *Systematic Biology*, 64(1):e42–e62, 2015. doi:10.1093/sysbio/syu048.

Simon Tavaré. Some probabilistic and statistical problems in the analysis of DNA sequences. In Robert M. Miura, editor, *Some Mathematical Questions in Biology—DNA Sequence Analysis*, volume 17 of *Lectures on Mathematics in the Life Sciences*, pages 57–86. American Mathematical Society, Providence, RI, 1986.

Nathan S Upham, Jacob A Esselstyn, and Walter Jetz. Inferring the mammal tree: Species-level sets of phylogenies for questions in ecology, evolution, and conservation. *PLoS Biology*, 17(12):e3000494, 2019. doi:10.1371/journal.pbio.3000494.

Josef C. Uyeda, Rosana Zenil-Ferguson, and Matthew W. Pennell. Rethinking phylogenetic comparative methods. *Systematic Biology*, 67(6):1091–1109, 2018. doi:10.1093/sysbio/syy031.

Günter P. Wagner. *Homology, Genes, and Evolutionary Innovation.* Princeton, 2014. doi:10.23943/princeton/9780691156460.001.0001.

Douglas B. West. *Introduction to Graph Theory.* Prentice Hall, 2nd edition, 2001.

April M. Wright, David W. Bapst, Joëlle Barido-Sottani, and Rachel C. M. Warnock. Integrating fossil observations into phylogenetics using the fossilized birth–death model. *Annual Review of Ecology, Evolution, and Systematics*, 53(1):251–273, 2022. doi:10.1146/annurev-ecolsys-102220-030855.

Yihui Xie. *bookdown: Authoring Books and Technical Documents with R Markdown.* Chapman and Hall/CRC, Boca Raton, Florida, 2016. ISBN 978-1138700109. URL https://bookdown.org/yihui/bookdown.

Ziheng Yang. Maximum likelihood phylogenetic estimation from DNA sequences with variable rates over sites: approximate methods. *Journal of Molecular Evolution*, 39(3):306–314, 1994. doi:10.1007/BF00160154.

Ziheng Yang. *Molecular Evolution: A Statistical Approach.* Oxford University Press, 2014.

Ziheng Yang and Bruce Rannala. Bayesian phylogenetic inference using DNA sequences: a Markov Chain Monte Carlo method. *Molecular Biology and Evolution*, 14(7):717–724, 1997. doi:10.1093/oxfordjournals.molbev.a025811.

Guangchuang Yu. *Data Integration, Manipulation and Visualization of Phylogenetic Trees.* Chapman; Hall/CRC, 2022. doi:10.1201/9781003279242.

Guangchuang Yu, Tommy Tsan-Yuk Lam, and Shuangbin Xu. *ggtree: an R package for visualization of tree and annotation data*, 2026. URL https://bioconductor.org/packages/ggtree. R package version 4.0.5.

Felipe Zapata, Freya E Goetz, Stephen A Smith, Mark Howison, Stefan Siebert, Samuel H Church, Steven M Sanders, Cheryl Lewis Ames, Catherine S McFadden, Scott C France, et al. Phylogenomic analyses support traditional relationships within Cnidaria. *PLoS ONE*, 10(10): e0139068, 2015. doi:10.1371/journal.pone.0139068.

Chao Zhang, Maryam Rabiee, Erfan Sayyari, and Siavash Mirarab. ASTRAL-III: polynomial time species tree reconstruction from partially resolved gene trees. *BMC Bioinformatics*, 19(S6):153, 2018. doi:10.1186/s12859-018-2129-y.

Chao Zhang, Celine Scornavacca, Erin K. Molloy, and Siavash Mirarab. ASTRAL-Pro: quartet-based species-tree inference despite paralogy. *Molecular Biology and Evolution*, 37(11):3292–3307, 2020. doi:10.1093/molbev/msaa139.

Jianzhi Zhang. Performance of likelihood ratio tests of evolutionary hypotheses under inadequate substitution models. *Molecular Biology and Evolution*, 16(6):868–875, 1999. doi:10.1093/oxfordjournals.molbev.a026171.

Emile Zuckerkandl and Linus Pauling. Molecules as documents of evolutionary history. *Journal of Theoretical Biology*, 8(2):357–366, 1965. doi:10.1016/0022-5193(65)90083-4.

Index

Akaike information criterion (AIC), 79
amino acid data, 103
amino acid substitution model, 103
ancestral state reconstruction, 115

Bayes' theorem, 94
Bayesian information criterion (BIC), 79
bootstrap, 83
branch length, 15
Brownian motion, 105
burn-in, 97

chronogram, 15
clade, 17
cladogram, 109
coalescence, 86

discordance, 85

effective population size, 86

fossil calibration, 112

gene family, 89
gene tree, 85

homology, 3
horizontal gene transfer, 91
hybridization, 91

incomplete lineage sorting, 86
independent contrasts, 126
ingroup, 22
introgression, 91

likelihood, 55
likelihood ratio test, 78

Markov chain Monte Carlo (MCMC), 96
maximum likelihood, 62
Mk model, 104
molecular clock, 111
  relaxed, 111
monophyletic group, 18
multispecies coalescent, 86

Newick format, 26
node
  internal, 12
  tip, 11

optimality criterion, 63
Ornstein-Uhlenbeck model, 107
orthology, 89
overfitting, 79

paralogy, 89
paraphyletic group, 19
parsimony, 63
phylogenetic generalized least squares (PGLS), 127
phylogenetic network, 91
phylogenetic non-independence, 121
phylogenetic signal, 128
phylogram, 109
polyphyletic group, 19
polytomy, 19
posterior probability, 95
prior probability, 94, 95

RAD-seq, 69

reconciliation, 90
root, 19
rooting, 19
    midpoint, 22
    outgroup, 22

sequence alignment, 73
sister group, 27
species tree, 85
substitution model, 47
    General Time Reversible (GTR), 48
    Jukes-Cantor (JC69), 47

targeted enrichment, 68
taxon sampling, 65
time calibration, 111
topology, 24
topology test, 83
transcriptome, 68

ultrametric tree, 15

www.ingramcontent.com/pod-product-compliance
Lightning Source LLC
LaVergne TN
LVHW070522061126
851857LV00023B/2128
* 9 7 9 8 9 9 3 4 5 2 4 0 1 *